The Complete Guide to Security

Martin Clifford was born in Harlem in Manhattan and is a graduate of the City University of New York. Subsequently he worked as an engineer for Sperry Gyroscope Co. with emphasis on the areas of voltage regulated power supplies, radar, and automatic direction finders for aircraft. He also acquired various licenses—radio amateur (ex W2CDV), first-class radio-telephone (commercial), and a license to teach vocational subjects from the University of the State of New York.

When the demand for engineers grew slack he became a teacher in private vocational schools and specialized in electronic communications. During this time he became interested in writing and gradually became a full-time free lance writer and has had numerous articles published in newspapers, trade, and consumer magazines. His articles have been syndicated in as many as 200 college newspapers.

Most of his writing efforts are devoted to books. He is the author of many books as well as a number of booklets on various topics in electronics, and several correspondence courses in electronics, radio, television, and drafting. He is the author of *The Complete Guide to Car Audio* and *The Complete Guide to High Fidelity.*

The Complete Guide to Security

by Martin Clifford

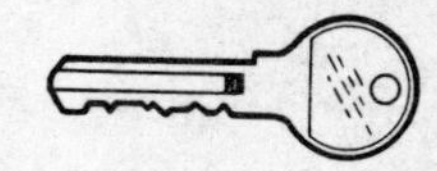

Howard W. Sams & Co., Inc.
4300 WEST 62ND ST. INDIANAPOLIS, INDIANA 46268 USA

FIRST EDITION
FIRST PRINTING—1982

International Standard Book Number: 0-672-21955-7
Library of Congress Catalog Card Number: 82-50656

Edited by: *Jim Rounds*
Illustrated by: *T. R. Emrick*
Kevin Caddell

Printed in the United States of America.

Preface

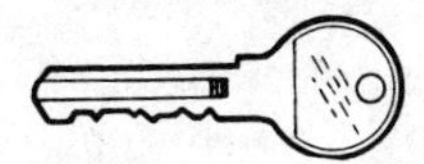

Crime is big business and is one of the largest enterprises in our country, probably because it is so pervasive. It can consist of just the few dimes or dollars taken from school children at knifepoint but it can also be the theft of millions. Crime involves everyone. There are no exceptions, for even if you aren't the direct victim of a thief you must still pay. You pay for police protection and if more police are needed, and they are, you pay in the form of increased taxes. You pay if you buy crime prevention units for your home, for if there were no possibility of burglary you would not need to make such purchases.

You pay when you buy a lock for your door or when you get a replacement lock. You pay when you buy insurance, whether for your home or car, and often your costs here are high. If you are injured as the result of an assault you pay, not only in the form of medical bills, but in physical suffering and mental anguish.

Crime affects all of us and there are no exceptions. It includes the young and the not so young. It affects men and women; even children are no exceptions. There is no way in which you can exclude yourself, no way in which someone else can be made to bear the burden of crime. We all bear it, we all pay, not only in money but in fear, in loss of peace of mind. Crime often dictates where you may or may not live. A high crime area might be more conveniently located near your place of employment but you may not be able to consider it.

Crime forces you to think of an unpleasant subject when you would rather concentrate on your life, your work, your future.

Crime makes demands on your consciousness, for one of our possible weapons against crime is a constant awareness of it.

Crime is counterproductive. It adds nothing to our gross national product. It increases your tax burden, for criminals are not tax payers. And it may seem to you that laws are made for the protection of the criminal. It may be difficult for you to accept, but the fact is your rights are being protected at the same time. And, while you may be the victim, it is often difficult, if not impossible, to determine the extent to which you cooperated in making the crime possible.

There is a prevailing attitude that more attention, more legal protection is given to the criminal rather than his victim. This concept is bringing about some changes. In one state, Missouri, a recent law fines convicted persons with most of the funds used as compensation for victims. The victim of a crime can collect up to $10,000 for hospital bills not recoverable from insurance, or for loss of income. Some of the other states are beginning to follow this concept.

Many of the topics discussed in the first chapter are covered in detail elsewhere. You will also find similar security measures described, not in just one chapter, but in several. It isn't always easy to put crime into nice, neat little packages. Robbery may be accompanied by rape. Burglary may also involve vandalism. Some security measures used in a home or business are also applicable to cars. And so, in reading the different chapters, topics covered earlier may be described again but in another context and possibly in greater detail. Use the contents pages and the index as a convenient cross referencing guide.

There are numerous manufacturers of anticrime products and it would be impossible to mention them all. A few are given as representative and you will find a list of manufacturers and their addresses at the back of this book. The list isn't complete nor was it intended that it should be. A variety of anticrime products is usually available in hardware stores, home supply stores, and department stores. The descriptions supplied in the various chapters are just intended to give you some conception of the different types and the pros and cons of each.

MARTIN CLIFFORD

Contents

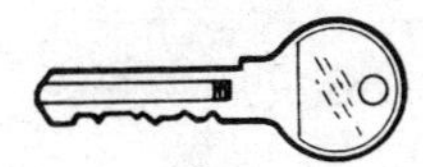

Chapter 1

The Need To Fight Back

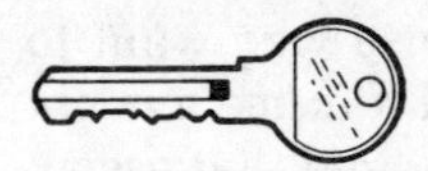

It is difficult to be constantly conscious of the possibility of a crime against you for a number of reasons, even though such awareness is a crime preventive. Like anyone else, you are accustomed to concentrating on your personal affairs and so are mentally preoccupied. Further, you have also learned to rely on others, such as the police, for your protection. Finally, you may think of a crime as something that only happens to someone else, but not to you, or you may consider crime as something that takes place only in a dark alley or on a deserted street. Surrounded by people, in broad daylight, and in a shopping plaza you have used many times without incident, you feel confident and secure. But the incidence of crimes in such hitherto secure locations is increasing. Your car can be stolen from a shopping plaza, or you can be robbed. Women have been abducted from such areas, raped, and robbed.

BE ALERT

Your first line of defense is to be alert. Although you may not be aware of it, no thief attacks a victim without first making an assessment. Are your hands loaded with packages? Do you seem to be preoccupied? Are you alone? Are you disabled in any way or are you a senior citizen? Are you carrying your purse so it can be easily snatched? The same considerations apply to men, although women are more likely to become victims of an assault. One thing the thief looks for is an easy mark.

ABSOLUTE SECURITY

There is no such thing as absolute security since we must work with, live with, and associate with people. Man is a social animal, but sometimes he is a bit more animal than social. There are many things you can do to protect your person and your property, and interestingly, these many things do not always involve the purchase and installation of protective equipment.

Crime is like death. We are, all of us, reluctant to admit it can happen to us. And, yet, it cannot only happen, but often does happen because we are so cooperative with those who want to rob, maim, or kill us. To be as free as possible of a crime against you means you must take an important first step, and that essential step is *awareness*. Call it crime-consciousness, if you like, or alertness, or anything else. The fact is that if you have a realization of the possibility of crime, then you have taken a forward move to eliminate or to minimize it.

THE NEED FOR AWARENESS

Many crime-prevention techniques are based on such awareness. If you are going to park your car on the street, select, if possible, a location under a bright street lamp. Close the windows of the car and make sure the doors are locked. Don't take for granted that slamming the door of a car will automatically lock it. It may or may not. Try the doors—not just the one you used, but all of them. Don't leave anything of value out on the seat where it can be seen; and don't try to cover merchandise or personal items with a blanket or newspapers. These possessions, plus any suitcases, should be stored in the trunk, and smaller items in the glove compartment. Valuable electronic equipment such as a CB unit, stereo system, etc., are fair game for any thief who spots them through your car window. Use an auto alarm and make sure it is set before you leave the car. Affix a side window sticker emphasizing that your car is burglar-proofed.

Will all of these steps guarantee your car against theft?

No, they won't do that, but they will increase the odds in your favor, and that is about the best you can hope for. The important point here is that awareness has made you less susceptible to becoming still one more crime statistic.

This concept of awareness applies to everything you own—your car, your home, and more important, to your own person. If you must walk down a deserted street, follow the most lighted path; keep away from doorways and empty lots. If you must walk through a high crime area in the daytime, even with people out on the streets, take certain precautions before you do so. Don't keep your wallet in your hip pocket or a pants pocket. Put it in your jacket breast pocket. And make sure that breast pocket is a deep one. When you buy a suit, check the breast pocket to make sure it extends at least one inch—preferably more—above the top level of your wallet. If not, have your tailor alter the pocket. Avoid small wallets that form a bulge. A bulge in your breast pocket is an advertisement.

Of course, these are all common-sense suggestions, but then common sense isn't quite as widespread as you might imagine. All of these things—avoiding a bulging pocket, locking car doors, and parking under lights—are simply a part of awareness. Your safety and security are dependent in great part on the constant and conscious effort you make to protect yourself and your property.

WALLETS AND POCKETBOOKS

For some thieves, anything you have and anything that can be resold are suitable targets. Wallets and pocketbooks are preferred since these may contain cash and credit cards. They may also hold house keys and car keys, plus personal identification. With your keys and your address, it doesn't take a very imaginative thief to realize he has entree to your home and your car.

If you use credit cards, keep a record of the account numbers at home and notify the credit-card company immediately if you lose your cards.

Obviously, you are going to carry some money and identification in your wallet or pocketbook. There are two things you must do as far as pocketbooks are concerned: (1) Make it as difficult as possible for a pickpocket to get into it; and (2) make it as difficult as possible for someone to snatch the pocketbook from you. Avoid pocketbooks that depend only on friction closing. If you can open the pocketbook with a flick of your fingers, so can a thief. Get a pocketbook with positive-locking type closure, preferably one that requires a sliding movement in two directions. That isn't all, though. That pocketbook should have a pocket with a zipper, and the longer the zipper the better. Inside that zippered pocket you should have a change purse, but make sure that the change purse has a lock that can't be opened with one finger. Now, that's a lot of trouble and inconvenience for you when you want to get at your money, but if it is difficult for you (and you are familiar with your pocketbook) think of how much more troublesome it will be for a pickpocket.

Snatch and Run

Not all thieves have pickpocket skill. Some of them depend on snatch and run. This involves a number of stealing techniques. Some thieves prefer cutting the shoulder straps of your bag, and they can do this and be off and running with it before you have recovered from your shock long enough to scream. The solution is to make sure the shoulder strap is as strong as possible. Some leatherette straps are so thin and flimsy they can be cut with a cuticle scissor. A metal pocketbook strap is best; if unavailable, get one that is multi-stranded.

The safest way to carry a pocketbook is with the strap on the shoulder opposite the pocketbook. This means your head and part of your body are covered by the straps. If you carry your purse in your hand, it can be snatched with a force and vigor that will leave you speechless. If you must carry your pocketbook, at least twist the strap around your arm several times. You can be sure the thief isn't going to snatch just any purse. He is going to take the line of least resistance and make off with the pocketbook that offers the greatest chance of success.

A wallet has a better chance for survival, and as mentioned earlier, is best carried in a breast pocket. An easy form of protection is to use a zipper across the top of the pocket. The zipper cannot be seen and does not affect the wearability of the garment. For best security, do not use the "straight-across" kind of zipper. Instead, get one that goes across and then turns down at a right angle for about one inch or so. You can, as a temporary expedient, use a safety pin. Two pins are better. Yes, it does sound silly to tie yourself up with "diaper pins," but losing your wallet to a pickpocket does indicate he is more security-conscious and alert than you are.

Hat Theft

Ladies' hats, and men's hatwear, particularly the more expensive types, can be and are stolen. A mink hat is a tempting item. The only way you can protect expensive headgear is not to wear it in very crowded areas, or in high-crime areas. This doesn't mean the hat won't be stolen—just that you have become security-conscious and are trying to improve the odds somewhat.

Jewelry Theft

Next to wallets and pocketbooks, jewelry offers an enticing target. If you must wear a diamond ring, make sure it fits and that you wear guard rings with it. Wear valuable pendants, earrings, rings, brooches, or pins only in safe areas. No restaurant is a safe area, no matter how well it is lighted or attended. No theater is a safe area. No public place is a safe area. Any excessive display of jewelry attracts attention, and once that happens you may be selected and followed. Security means being conscious of possible trouble, anticipating it, and doing everything you can to avoid it.

There is no place where security isn't important. If you are on a subway, select that car used by the conductor for opening and closing the doors. This means you should avoid the first and last cars, and stay somewhere toward the center. If you must go on a subway or train platform at night, don't do so unless the platform is occupied by people. Stay where people are. Avoid subway and train passageways that are empty and dimly lit. Try to schedule your traveling so that you avoid very late night hours, if possible. Bus and train terminals can be areas of personal danger during those times when "people traffic" is light or practically nonexistent. If your work schedule is such that it involves night travel, consider whether your life and property are worth the risk you must take.

How can you protect yourself when you are away from your home or office? Unless you are licensed to carry a gun, and know how to use one, carrying a gun offers no protection. If you are faced with a knife and/or gun, surrender your valuables as quietly and as quickly as possible. Being robbed is insulting, degrading, and demeaning, to say nothing of the fact that you may have worked hard for your money, and that the financial loss may cause you worry and deprivation. No matter. Loss of life or a serious injury is worse.

If you must walk alone and you do want protection, you can get some measure of it from a dog. Not any dog, by any means, but one that is trained to respond to your command. The ordinary household pet can be dragooned into watchdog services, but classifying all dogs as the same is about as sensible as classifying all people as the same. Your dog may very well be more afraid of strangers than you are. If you do get a trained dog, make sure you follow the trainer's suggestions to the letter. A dog can be spoiled, just as a child or an adult can be spoiled.

Mace/Police Whistle

A chemical spray such as Mace or a police whistle may be helpful, depending entirely on you and also on the circumstances in which you find yourself. You can be sure the person trying to rob you isn't going to stand around and wait while you rummage around in your pocketbook for your chemical spray. Consider also that the thief may take the Mace away from you and use it against you, or you may be spraying yourself if the wind is blowing in the wrong direction.

A whistle is somewhat better but, by the time you recover from your shock, you will probably find that your assailant has long since gone. The best method, then, is anticipation and try to avoid situations that will make you a victim. Stay in lighted places, walk or travel when others do. You are a likely victim if you enter your elevator alone with your arms loaded with packages. And don't invite strangers to your apartment. Your intentions may be good, but you may be the only one having such intentions.

FACING A THIEF

If you ever feel a gun shoved into your back, or find yourself facing one, don't try to be a hero. You may go the rest of your life and never be held up at gunpoint. Or, it could happen tomorrow. There are no hard and fast rules to go by should you suddenly be looking down the barrel of a gun or feel the thrust of cold steel against your back. Most authorities agree, however, that if this does happen to you, it is wise to yield to the demands of the gunman rather than to attempt to counteract with some sudden movement of your own.

Since most armed robbers are after money or other valuables, it is sensible to do exactly as you are told, quietly and carefully. This may be difficult under traumatic circumstances such as these, but you can gain some degree of comfort, should it happen to you, in the knowledge that your assailant could be as frightened as you are, and wants to leave you and the scene of the crime as rapidly as possible.

But isn't there anything you can do, without risk to yourself?

Yes, there is, and that is to keep your wits and use your powers of observation to the fullest extent. Note the height, weight, age, and coloring of the criminal. Look for distinguishing features or characteristics. Try to remember color of hair, color of eyes, marks on the face—anything that will help the police. Try to remember the sound of the voice. Observe the kind of clothing your assailant is wearing, or anything at all that is unusual about him. If a car is used, make a mental note of the license number, or if not, at least something about the car that would be helpful in identifying it. And then report your mishap to the police as quickly as possible.

It is easy to give advice, and much more difficult to be the victim of a crime. Being a crime victim and emotional shock go hand in hand, but if you want to fight back, then this is possibly the best way you can do it.

While you may not be able to prevent a robbery, you may be able to minimize the possibility by taking evasive action or by reducing the consequences. If, for example, you must deposit a large sum of money in the bank every day, increase the number of trips you take, and reduce the amount you carry. Don't ever carry more cash with you than you really need. And some people are so credit-card happy that they may carry a dozen or more of them at all times. A credit card is more than cash, for it is an unlimited demand on your resources. Restrict your credit cards to those that are absolutely essential. Possession of a large number of credit cards is regarded as a status symbol by some, but it is a symbol that carries high personal risk.

How To Minimize Your Risk on the Street

A street criminal may be a loner, may form a partnership with another criminal, or may be part of a gang. But no matter how these criminals work, whether individually or not, they are opportunists. What they prefer is to operate against a single person, in a deserted area. This doesn't necessarily mean you are safe in a well-lighted, crowded area. A mugger will look for lone victims in out-of-the-way places. Pickpockets prefer crowded, busy locations and people who are careless about their wallets or pocketbooks. Purse snatchers and jewelry snatchers, most of whom are young people, also like crowded areas into which they can disappear. They often select women as their victims since the woman's possessions, her cash and credit cards, are out in the open in a purse, often loosely held.

A street criminal has a definite advantage. Such a person is organized, you are not. Your thoughts and concentration are on a number of things that do not include crime. The thief is organized in the sense that there is just one thought and one objective. There is also the advantage of the element of surprise and that of prior planning. But you can make it more difficult for the criminal by following a few rules.

For women who carry purses, and almost all of them do, it would be advisable to have a separate purse or wallet for money, credit cards, and house keys. These can be kept in an inside pocket and it would be advisable to have such a pocket stitched into a coat, if it isn't so equipped. If you must carry a purse, always make sure it is securely closed. Use a purse that locks securely. A shoulder strap purse is better than a hand-held type, provided you can slip the strap of the purse over your head. A hand-held purse offers the least security and one that dangles from a shoulder is almost as bad.

Try to avoid showing any more cash that you absolutely must. If you use the separate wallet and purse arrangement, and you should, keep no more money in your purse than you need. Do not carry large denomination bills, such as $50 and $100. It takes time to get change for such bills and during that time a thief could be estimating his chances of separating you from your money.

Many streets in downtown business districts are deserted at night. Try to arrange your schedule so you need not walk on such streets. Avoid shrubbery, parked cars, empty lots, and streets that are poorly lit. Avoid parked cars. If a car follows you, immediately turn and walk in the opposite direction for you can make the turn more quickly than a car. If at all possible, try to have one or more, preferably more, friends accompany you, especially in high risk neighborhoods.

Do not wait alone at deserted bus stops. If it is feasible, select a bus stop used by other people, even if such a stop isn't convenient. If you must use a particular bus stop, try to get the bus company to change that stop to a more preferred location. You will have a better chance at success if you do this through a group, rather than as an individual.

It is generally not advisable to resist during an attempted robbery, if you see or suspect that your attacker can use force.

You could be seriously hurt. You can always replace cash, jewelry, or a credit card; and you can do this much more easily and with less pain than waiting for your body to heal. During an attack, sit down if you can, even if it means sitting on the sidewalk. This is better than being knocked down. Don't try to follow your attacker. Instead, call the police as soon as you possibly can.

While walking, look ahead from time to time. Don't window shop if the street is deserted or has few pedestrians. If you are suspicious of persons approaching you, cross to the other side of the street. The advantage of this is that it shows you are alert to the possibility of theft, something no attacker wants you to be for attackers often depend on the element of surprise.

If you must carry credit cards, make a written record of each card identification number and keep that information at home so you can use it in the event of loss, or theft. If you do lose one or more credit cards, notify the issuer immediately. Along with credit card information, keep the telephone number of the issuing company so that you do not waste time looking for it.

LAUNDRY ROOM AND LAUNDROMAT

The laundry room of an apartment house and the private store laundromat are possible scenes of crimes. The thief knows your hands will be occupied and the area sufficiently private for an assault.

There are several common-sense rules to follow. For a laundry room, try to arrange to do your laundry with at least one other person, but two or more are better. The same applies to a laundromat. Do not enter it, if, by looking through the window you can see you will be alone.

Some laundromats are always attended by the owner, or by an employee. Patronize those that have attendants instead of coin-operated laundromats which can be completely unoccupied.

SCHOOLS

Do you live near a high school? Young offenders often begin as purse snatchers. And if you are alone and a senior citizen, you can avoid accommodating a thief by staying away from these schools at the start of the school day, during recess or lunch periods, or when the school is being dismissed.

Crime isn't something that involves adults only. During 1980, New York City Police arrested 12,762 children, age 16 and under, on felony charges. In Chicago, the figure for the same period was 18,754 charges. Much of teenage crime is committed by middle-class youngsters.

In Los Angeles, police estimate that gang activity accounts for one out of five murders and robberies.

CRIMES AGAINST ELDERLY PEOPLE

It is commonly thought that elderly people, our senior citizens, are the most common victims of crime, possibly because of the publicity such crimes have in the media. However, senior citizens are less likely to be the victims of crime than people who are younger. In a study conducted by the Department of Justice, statistics indicated that persons 25 years of age and younger were most often the victims and this includes individuals whose age is as low as 12.

This doesn't mean that senior citizens are immune against crime, for they are choice victims for purse snatchers and holdup men. Often, the victim is the target of a group, two or more; and equally often the group will follow the victim for an opportune moment or place in which to make an attack. About 75 percent of all thefts committed against the elderly are common thefts.

The elderly are the major victims in cases of personal larceny involving contact between the thief and an individual. Personal larceny, in this case, is the theft of a purse, wallet, cash, and may even consist of packages being carried by the victim. Still, crimes of violence against people 65, or older, decreased from 8.5 per 1,000 in 1973 to 6.9 per thousand in 1980.

It is difficult to determine the precise reason for this drop in theft, but it may be due to increased caution on the part of senior citizens and to the fact they may have become more security conscious. Thus, senior citizens may go shopping in groups or may be accompanied by stronger, younger people hired for protection. They may also have made themselves less inviting targets by leaving jewelry and extra cash at home.

If you are a senior citizen, one of the first steps toward protection is to learn just what your risks are. Speak to your neighbors to learn more about the incidence of crime in your neighborhood. Also take the time to visit your local police department. They may have a counselor for groups of senior citizens.

Surveys by the Federal Government show that seniors are less likely to be the victims of the most serious violent crimes, including murder, rape, and aggravated assault. This does not mean it cannot happen, just that it is less likely.

The same survey shows, though, that older people are more likely to be the victims of certain crimes, including purse snatching, mugging, theft of checks from the mail, burglary, and fraud.

The survey also shows that the risks are higher for those older persons who live in large urban areas than for those in rural locations. Crime is more likely in regions having a high population density and these are sections of the city in which personal income is low. However, the ratio of crime in the cities compared to crime in the suburbs and country is changing since crime in these places is rising.

Elderly people do not move from high crime areas for a number of reasons. They may not be able to afford to do so. They may have lived in a community for many years, have friends in that community, and are reluctant to start a new life elsewhere. And, because of their age, they may not have the spirit of adventure that would enable them to move.

For senior citizens, a good approach to protection is through a local neighborhood association or senior citizen program. Seniors are most likely to get added police protection if they approach the police as an organized group. In some areas, the police can supply an officer who will help initiate a self-protection program and will supply crime-prevention tips.

WORKING WITH THE POLICE

A police officer is a professional, someone who has usually received Police Academy training, supplemented by on-the-job experience. He or she is someone who has learned how to use a gun and how to apprehend criminals. The officer is paid to protect you and your property. But, like anyone else, he or she must work within the law. No one has ever estimated the number of criminals there are to every one police officer, but it is no exaggeration to state that the police are far outnumbered.

It would seem that the police would eagerly welcome help from concerned citizens, but it is a fact that a citizens group has no training, can easily violate laws, can easily interfere with police activities, and could possibly become a vigilante group.

It is always best, when forming a self-protection group, to work with the police and a good first step would be to contact them and consult them before becoming involved in any self-protection activities.

Concerned citizens groups operate in various ways. One method is to patrol their neighborhood in cars equipped with spotlights and CB radios; another is to cover the same area on foot, keeping a lookout for suspicious vehicles and people. Most groups simply instruct their members and neighbors to maintain a watch from their windows. In all cases, no matter which method is used, any unusual or suspicious activity should be reported immediately to the police. All it takes is a telephone call.

Another technique is to put signs on windows, on posts, on billboards, or on store fronts, that the area is under surveillance by alert citizens and by unmarked cars.

Helping the Police

A growing number of police and sheriff's departments across the country recognize and use the talent and energy of America's retired persons.

In one area (Far Rockaway, N.Y.), senior volunteers monitor a base-station CB radio located at the police station. When a call for police service is received via CB, they relay it to the police dispatcher. They have helped with calls about crimes in progress, auto accidents, boats in distress, fires, plane crashes, and other emergency situations.

In Sun City, 20 miles from Phoenix, AR, a retirement community has formed a large volunteer posse, working with the county sheriff's department. Groups of senior citizens patrol their neighborhoods on foot, bicycles, and even in golf carts. They also go door-to-door to promote operation ID and neighborhood watch. By doing these things they free more officers for patrols. The sheriff's department estimates that these activities save taxpayers about one million dollars a year.

In San Diego's police department, a dozen senior aides help by encoding data on incident reports, entering and checking computer data on reports, and studying false burglary alarm reports. The AARP (American Association of Retired Persons) and NRTA (National Retired Teachers Association) have jointly developed a package of training materials for interested police or sheriff's departments to use in starting a crime analysis unit and then to place retired or older volunteers and/or paid assistants in crime analysis support roles.

Ask your local law enforcement agency or senior citizens' center about crime-prevention programs in your community. If none exist, organize your own group to take action against crime.

Work With Young People

Nearly half the violent crimes in cities are committed by young people, 21 and under. There are many opportunities for senior citizens to work on a one-to-one basis with young people. Some work with teenagers who are thinking of dropping out of school; in other communities volunteers are encouraged to help as teaching assistants, counselors, or special tutors.

ACTION

A federal agency, Action, has set up several programs for older Americans. The Retired Senior Volunteer Program (RSVP) which operates in over 700 communities, provides volunteer assignments for anyone 60 or older who wants to serve others in the community. Action also sponsors the foster grandparent program, a part-time opportunity for low income senior citizens who want to provide much needed attention to disadvantaged youngsters in their community, based on the concept that help supplied now could prevent a crime later.

AID THE COURTS

Working with the police is just one of the ways that an individual can help in the fight against crime. Learn about your court system and with the support of your local police, prosecutors, and court officials, help improve the treatment of witnesses and victims, and the efficiency of the court.

In Milwaukee, WI, 15 senior volunteers with the encouragement of the district attorney, follow all criminal cases involving older victims or witnesses to determine if they are being treated fairly and considerately in the process. In addition, volunteers contact the victims or witnesses prior to their court appearances, greet them upon their arrival at the court, and accompany them throughout the entire court procedure. The volunteers' observations are recorded and the results are reported periodically to the county government, the office on aging, and various law enforcement and government agencies.

In another area, volunteers from the Retired Senior Volunteer Program (RSVP) have succeeded in bringing order out of chaos in the busy waiting room of the juvenile court which handles over 12,000 cases per year. Working in pairs, they run a reception desk in the court's waiting room where they register and keep track of all the persons who arrived for each case, explaining procedures and when and where the cases will take place. Their presence is a calming influence on nervous citizens called to the courts and helps reduce the stress on court personnel.

NEIGHBORHOOD ACTION

There are two important roads to security. One is to make your car, your home, and other possessions as free from ripoff as possible. The other is to form a group with friends and neighbors with the explicit purpose of self-protection. As a start, visit your local police department or sheriff's office to learn if such a group is already in existence, and if not, find out how to go about forming one. There are a number of ways and the following consists of suggestions based on groups now at work.

Security Inspection

Many people aren't aware of the need for security and become conscious of it only after a robbery, a mugging, or ripoff. In Cottage Grove, OR, six senior citizens have learned enough about neighborhood crime so that they can function as crime-prevention specialists. They share their acquired expertise and visit homes to instruct residents in security measures. They also serve as speakers at public functions to increase awareness of crime.

In St. Louis, MO, a group of 84 senior citizens, with prior instruction and training by police, do an average of 140 home-security inspections a month. Trained to detect security weaknesses, they help senior citizens. They do more than just advise, for they also install locks and peepholes and fix broken windows and light fixtures.

Operation Identification

In Natick, MA, a group of seniors has formed Operation Identification. These volunteers, equipped with electric pencils, put identifying numbers, such as Social-Security numbers, on valuable equipment—on typewriters, on stereo equipment, and other items. They also supply Operation Identification stickers to post in windows as a warning to potential burglars. The identification process is an aid to police in recovering stolen property and in returning it to the rightful owners.

The Observers

Not all people are sufficiently active or capable of becoming involved in security activities directly, and many are often fearful of becoming involved. They can still help by working as silent observers. They watch and observe, reporting suspicious events.

In Mansfield, OH, police issue confidential identifying numbers to a group of senior citizens who agree to accept observer status. They are able to report unusual activities and can do so in complete confidence, without fear of reprisals. Their names are not revealed at any time.

In Battle Creek, MI, a group of such anonymous observers have reported crimes and have received cash awards for valuable information. They have now been functioning for over 12 years.

Neighborhood Watch

Like the Observers, a group of neighbors in an apartment house or in homes near each other can organize to form a neighborhood watch. They, too, can detect any unusual activities. The neighborhood watch program can be set up so that each member has a daily schedule. Telephone numbers are exchanged so they can remain in communication with each other. A neighborhood watch is often more difficult to set up in an apartment house, since the tendency of apartment residents is to be somewhat reclusive. It's an educational process and it does take some attitude changes.

Buddy Buzzer System

In a retirement building in New York, some tenants have installed intercommunicating buzzers that sound in one or two other apartments. A call for help involves nothing more than pressing the button on the buzzer. This has an advantage over a telephone since no dialing is required. The sound of the buzzer is an indication that help is needed.

There are some disadvantages to this technique, though. Permission is required from the owners of the building. Not more than two or three apartments can be connected since the buzzer doesn't give any indication as to which apartment requires help.

Further, the residents receiving a buzzer warning may not be at home at the time.

Citizen Escort

In Wilmington, DE, young people and other volunteers accompany senior citizens on their errands. On the lower east side of New York City there are ten minibuses that bring senior citizens to health-care facilities, shopping areas, and other functions. This is done without charge. A program of this kind requires financing, either from a municipality, or from companies and/or individuals willing to make contributions.

Citizens' Patrol

A patrol can consist of two or more persons and the controlled area can be limited to a group of apartment houses, or it may cover an entire neighborhood. In Wilmington, police have provided radio monitors to selected senior citizens in high-rise apartments. If a crime occurs, about 15 seniors are alerted. The seniors are not expected to follow suspects, merely to report suspicious persons running or hiding. The information is then telephoned to the police.

Important factors in a Citizens' Patrol, or in any other activity involving citizens, are organization and scheduling. A haphazard setup is of little value. A time schedule must be maintained, the activities of each person must be carefully explained, and the method of operation well understood by everyone participating.

Helping the Victims

To many it does seem as though our laws are designed for the protection of perpetrators rather than for victims. That is the price we pay for having a democratic society in which a person is presumed innocent until proven guilty.

Unfortunately, victims often face hardships as a result of physical injury or loss of property. Volunteers are needed to help victims as much as they are needed in crime prevention.

In Pasadena, CA, senior citizens who are crime victims receive immediate attention and help from a group of trained senior volunteers. These volunteers who are available on a 24-hour basis supply victim assistance to about 85 older victims every month.

In Yonkers, NY, a Care Team comprising about 20 volunteers use a car donated by the police department to make visits to the homes of older victims who have asked for help following a crime. These volunteers also work with the police, by providing newcomers to the police force with information about the special needs of senior citizens.

MUGGING

Muggers use various robbery techniques and are variously known as yoke men, crib men, push-in robbers, and purse snatchers. A *yoke* man is the one who puts his arm around the throat of the victim, while the others, two, three or more, relieve the victim of his valuables. A *crib* job is one in which the criminals follow a selected victim to his (or her) door, pushing their way in as soon as the door is opened. It is known as a crib job since getting the money is as easy as taking things from a baby in its crib. A *push-in* robber is another name for a crib job. A *purse snatcher* is a criminal who prefers pulling a purse away from a woman.

Muggers are often young people, usually under 20 years. They are also responsible for burglaries and grand larcenies.

Muggers are often equipped with weapons, usually a knife. They not only rob the victim, but often commit acts of violence. If a mugging is inevitable, the only thing to do is to cooperate, but even that is no assurance of not being hurt.

HANDGUN CONTROL

FBI statistics show that about 50% of all murders in the U.S. involve the use of handguns. The subject of handgun control is a highly emotional one and you can hear excellent arguments in favor of handguns and an equal number of excellent arguments opposed to them. Municipalities have their own laws governing guns and these vary from one location to another.

FIRE CONTROL

People are not usually as concerned about fires as they are about security. Most homes aren't fire alarm equipped, although battery-operated alarms are quite inexpensive, are easy to install, and work effectively. Some even supply a warning sound when the battery needs replacement. The alarms are automatic and sound even in the presence of a small amount of smoke.

Homes with children are particularly in need of fire alarms since it is estimated that about 40% of all fires are set by children under the age of 6. Playing with matches is often a childish occupation and it isn't at all unusual for a child to start a fire. This is normal and can be attributed to curiosity.

However, there are also times when a fire is started as a way for a child to express his or her resentment, or for attracting attention. In families that are facing marital problems, you may find children, usually in the 5- to 8-year age bracket, setting a fire as a way of relieving tension and stress. Oddly, parents of children who set fires will often deny this activity, possibly due to shame or as recognition of their inability to solve their personal problems.

Fire setting on the part of children is not a problem that cannot be solved. Sometimes all that is needed is to determine the underlying cause, the motivation behind the need to set fires. Quite often it isn't even necessary to solve the problem, but just to give the child a chance to express himself. Outside counseling may be necessary, especially in cases where the parents refuse to admit that anything is wrong. Fortunately, most children limit themselves to setting only one fire.

WHERE TO GET MORE INFORMATION

You can get a booklet called Senior Citizens Against Crime, published by the Office of Justice Assistance Research and Statistics, U.S. Department of Justice. The publication is part of a series of 11 booklets produced by the Crime Prevention Coalition. This is a group of 48 public and private organizations that sponsors a national citizen crime-prevention publication called *Take a Bite Out of Crime.* For more information, write the Crime Prevention Coalition, P.O. Box 6600, Rockville, MD 20850.

For information on crime prevention and senior citizen volunteer efforts write to Criminal Justice Services, American Association of Retired Persons (AARP)/National Retired Teachers Association (NRTA), 1909 K St., NW, Washington, DC 20049. Or call (202) 872-4912.

For information on victim assistance programs and volunteer efforts, contact the Criminal Justice and the Elderly Program, National Council of Senior Citizens, 925 15th St., NW, Washington, DC 20005. They can also be reached by phone by calling (202) 347-8800.

While a number of organizations do specialize in helping seniors, there are some that help people of all ages. For information on how to set up a local program for the prevention of crime, write to

The Eisenhower Foundation for the Prevention of Violence
1666 K St., NW, Washington, DC 20006.

The Eisenhower Foundation has a toll free number you can use if you prefer to call. (800) 368-5664. Or, (202) 223-0530.

A group that will supply information on neighborhood patrol organizations is

National Center for Community Crime Prevention
P.O. Box 37456, Washington, DC 20013
(202) 783-6215.

For information about how to assist victims, contact either one or both of the following:

National Organization for Victims Assistance (NOVA)
918 16th St., NW, Washington, DC 20006
(202) 265-5042

Parents of Murdered Children
Charlotte and Bob Hullinger
1739 Bella Vista
Cincinnati, Ohio 45237
(513) 242-8025.

If you are interested in learning more about financial help for victims of crime, contact

The National Association of Criminal Victim Compensation Boards

Workmen's Compensation Bureau
Highway 83, N. Russell Bldg.
Bismark, ND 58505
(701) 224-2700.

If you are interested in learning more about gun control write to

Mr. Peter Shields
Handgun Control, Inc.
810 - 18th Street, NW
Washington, DC 20006

BURGLARY AND ROBBERY INSURANCE

The premiums to be paid for burglary and robbery insurance depend on the amount of coverage you want, your location, the number and kind of security installations you have, your record as an insurance risk, and the insurance company.

If you are unable to afford such insurance, consider Federal crime insurance, insurance that is sold by the U.S. government in 28 states.

The insurance will help you replace stolen possessions or repair any damage to property in a break-in, or both. It is low cost and provides up to $10,000 household protection. It cannot be cancelled regardless of the number or size of the claims you make. All you need to do to qualify is install approved locks on outside doors and windows.

For information and eligibility requirements, call any licensed property insurance agent or broker, or write to or call toll free: Federal Crime Insurance Program, P.O. Box 41033, Washington D.C. 20014, (800) 638-8780. In Washington, D.C. and suburban Maryland, call 652-2637; at other locations in Maryland you may call collect.

STATISTICS

Statistics about all types of crime are reportd on radio and television, and appear in newspapers, magazines, and books. While the numbers that are presented may appear impressive, their accuracy is suspect. All sorts of agencies report on crime at all levels of government, but there are no standards. A felony in one area could be a misdemeanor in another. Not all crimes are reported. A woman who has been raped or a man who has been the victim of a bunco scheme may be too ashamed to report the crime. Even if the criminal is caught, plea bargaining may reduce his crime from one level to another, but in the process crime statistics change.

This doesn't mean crime statistics are of no value. They do give us some measure of how well, or how poorly, we are doing in the fight against crime. All the numbers can do is to give us a rough approximation, and that is better than no estimate at all.

With these cautions in mind, we can consider Paris as a difficult city to live in for it has an average of 26 burglaries for every 1,000 residents, the highest burglary rate of any major city in the world, based on data supplied by the Center for Documentation of Insurance Information. Los Angeles follows closely with 25 burglaries per 1,000 residents; New York City, 22; London, 17; Chicago, 11; Brussels, 8; and Tokyo, 4. However, on a nationwide basis, the U.S. leads all the others with the highest burglary rate—15 per 1,000 inhabitants.

Chapter 2

How To Make Your Home More Secure

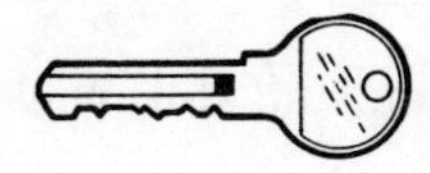

It is estimated that houses and apartments in the U.S. are burglarized on the average of once very 20 seconds. According to the Federal Bureau of Investigation (FBI) we can expect that one out of every four Americans will have their homes entered by burglars at some time in the future. Based on this data, you are more likely to be burglarized than to have an automobile accident or to become involved in a fire. Homes are also no exception to violence, since more than one in five violent crimes happen in the home.

Your home is where you live. It may be nothing more than a single room apartment or a private house with a dozen rooms. Your motel room or hotel room is your home for as long as you occupy it. If you rent a beach cabana it is your home for the short period of time you use it.

A home, then, isn't always a place you own, nor is it necessary for you to stay in it for years, months, or days. If you live, eat, and sleep in a trailer, then that is your home. But no matter what sort of residence you occupy, sooner or later—and often sooner—someone can get the idea that you represent a prospective victim.

WHY BURGLARY?

Burglary is a popular type of crime, with burglars that is. It can be a profitable crime, regardless of that time-worn cliche that insists crime does not pay. The risk for burglary is less than for other types of crime and it is made to order for the criminal who wants no physical involvement with his victims.

The attitude of many burglars is typified in the following excerpt from an actual interview of a young house burglar arrested in Salt Lake City, Utah, and now serving time in Utah State Prison:

Question: How many burglaries have you made in the last two years?

Answer: Close to one thousand.

Question: Did you just go for the rich home?

Answer: No, man, the poorest homes have tv sets and stuff you can move in a hurry.

Question: What about dogs?

Answer: They didn't really give me a problem. Dogs outside can be fed or put out. If a dog is in a house, he usually thinks you belong there.

Question: You'd go into a home with a dog inside?

Answer: Yes.

Question: Why do you think people are complacent about protection?

Answer: Most think they don't have anything worth stealing.

Question: Have you ever been in a home where you couldn't find anything worth stealing?

Answer: No. I've even taken a lamp table.

Question: What would you do if you got into a home and didn't find what you were expecting?

Answer: I'd bust up the joint, kick in the tv, and let off steam in some way.

Question: What if you walked into a room and an alarm sounded?

Answer: Man, I'd split.

Break-In Types

Burglary isn't the only driving motive for breaking and entering your home. A burglar-proof lock—that is, a lock that is as burglar-proof as you can get—will help hold out not just the professional (and also amateur) burglar, but other undesirable types as well. The vandal breaks into a home for "kicks" and resents anyone who has anything, including just a happy home. If discovered while breaking up a house, the vandal can be extremely dangerous.

The drug addict who has a fifty-to-one-hundred-dollar a day habit to support has only two ways to earn that kind of money—steal or push drugs. Most addicts steal because there is less risk. They burglarize many times while high and under these circumstances are very dangerous.

The sex pervert breaks into the home while someone is there for very obvious reasons.

Targets of Opportunity

Burglary is often the choice of crime by those who are inexperienced since it can be opportunistic. All the novice burglar needs is a house or apartment that appears to invite entry. And this is often supplied by the occupants when they are away and often enough when they are home.

No burglar is going to work any harder than he needs to. All a burglar requires are clues that your home is available, clues that may seem insignificant to you. Thieves look for what police call "targets of opportunity."

Who Are the Burglars?

There is a stereotyped conception of the burglar as a rough, uneducated individual, usually preferring force and dressed like a bum. Burglars are people and like other people can be so well dressed you could not possibly imagine their profession. Some are extremely well educated, highly cultured, amusing, witty, pleasant, courteous, and kindly. They are also crooks.

WHAT IS BURGLARY?

Burglary is a crime against a place or against property, not against people, or more appropriately, only against people indirectly. In other words, it is technically a structure that is victimized; although in common usage, we refer to the residents or owners of the structures as victims.

There is a difference between a burglar and the bunco operator, described later in Chapter 9. A burglar looks for likely places from which to steal, while the bunco or con artist looks for likely people to swindle. They both have the same objective. It is only their approach that is different.

Burglary, Robbery, and Larceny

There is a difference between burglary and robbery. A place cannot be robbed, but a person can be. Robbery is stealing from a person through the use of force or intimidation.

Taking property belonging to someone else is larceny. The laws that define robbery, burglary, and larceny vary from state to state, and in some, burglary is predicated on unlawful entry.

House Versus Apartment

Security for a house is often more difficult to achieve than in an apartment since there are so many more points of entry. However, if you own a house you can install whatever security devices you may think necessary. If you live in an apartment, or condominium, or in a retirement community, you may be restricted by your lease or by the rules of the condominium or retirement community as to what you may or may not do.

In any event, you and a group of other apartment owners or tenants can always speak to your building owners or operators about installing alarm systems, better lighting and locks, or closed-circuit television (cctv) cameras to watch hallways and entrances. In some buildings, cctv cameras are also installed in elevators for greater protection of the tenants.

Breaking In

Burglars often become aware of the ease with which entry can be made from clues left through the carelessness of victims. Doors and windows are often left unlocked, or in many cases even when they are locked, the locks are worthless and easily forced by the burglar with a plastic strip, or other simple tool. The burglar is often notified of the victim's absence from the premises through clues ranging from the obvious three-day accumulation of newspapers to the more subtle lone living room light shining away brightly at three o'clock in the morning. Local obituary columns and society pages can notify the burglar of places ripe for theft, as may comments made by potential victims or persons associated with these potential victims—servants, beauticians, or bartenders, for example—regarding clients' wealth and the occasions of their absence.

The burglar keeps up with the times and as a result steals more television sets than horses. Similarly, checks and credit cards have become more important targets of burglars in recent years.

The Burglar's Skills

The number and types of skills demonstrated by burglars are varied. They range from the relatively simple technique of throwing a rock through a window to gain entry, to the more complex use of lock-picking tools to overcome the barriers erected by the cautious property owner.

Similarly, a burglar's degree of skill is often shown in the types of goods he steals. The relatively unskilled or amateur burglar will generally seek money as his object of theft since this loot requires no knowledge of fences for disposal. On the other hand, the professional burglar has a wider number of contacts with receivers of stolen goods and the ability to distinguish between valuable and worthless items, a necessary ability in the case of furs or jewelry, for example. The burglar will often make this type of goods a prime target.

The highly skilled burglar is less common than the unskilled or semiskilled type, and the majority of burglaries that do occur are a result of opportunity: A thief sees promising circumstances—a window that looks partially open, a doorstep with an accummulation of newspapers, mail, and milk bottles—and takes advantage of them at that time. Some burglars develop what could be called the "larceny sense," the ability to sensitize themselves to a variety of illicit opportunities.

NIGHTTIME SECURITY

While the preference of a burglar is to ransack your home while you are away, many burglars have no objection to nighttime work for a number of reasons. They know an occupied home has many entry areas that are left unsecured. People often think an occupied home is a safe home, but that is just not the case.

The responsibility for nighttime security should be that of one person and one person only. Such security involves checking windows to make sure they are locked, to make sure all doors (front and rear) are secure, that the garage door is closed and locked, that all objects of value are inside the home and not out, and that any in-home alarm devices are in their on mode.

This doesn't mean all windows must be shut. In a two-story house the downstairs windows can be locked. Various sensor-type alarm systems can be used to cover window approaches. Also make sure all objects of value inside the home, such as wallets or purses, are put away. This doesn't mean a burglar won't find them, but there is no reason for making his work any easier.

An apartment is much easier to check than a home simply because there are fewer entrance areas. Still, it is a sensible precaution to make sure that the front door is securely locked. Windows that open onto a fire escape, or that are near it, should be locked. This does not mean they must be shut tight. There are window-locking devices that will let you keep windows partially open, but not open far enough to allow entry.

THE BURGLAR AND HIS VICTIM

A common misconception is that once a burglar is in a house one of his first steps will be to cut the power lines. On tv or in the movies, perhaps, but not as a real-life event. Many burglars are juveniles and they are literally afraid of power lines. Further, cutting a power line would make no sense since many alarm systems that are connected to an outlet are equipped with a battery backup system. If there is no line power, the battery takes over.

Burglary is basically a passive crime and one in which the burglar tries to avoid any form of contact with the victim. The reasons for this are varied. First, the chances of getting caught after committing a crime in an unoccupied house are lower because of the high probability that the burglar will be gone from the scene of the crime, and possibly rid of the stolen goods, before the burglary is discovered. Second, entering unoccupied premises has the advantage of minimizing the risk of later identification. Third, even if the burglar is caught, the penalties for this type of crime are likely to be less severe than those for other forms of theft, such as robbery. Fourth, the burglar is usually fearful of encountering his victim, realizing that such a meeting may endanger his own life as well as increase the risk of his apprehension and severe punishment.

Within the general category of burglary, there are a variety of choices the burglar can make. Some decide to be daytime burglars, working only during the day in vacant homes. Others prefer to steal at night; among them the so-called cat burglars, who like to enter while the victim is in, although characteristically do not generally seek a direct confrontation. Still other kinds include the hotel burglar, the jet-set burglar, and the apartment burglar.

Noise

The more highly skilled burglar may temporarily be dissuaded from committing a crime because of the unexpected occurrence of unforseen obstacles. A burglar will usually go ahead with a burglary once it has been decided to do it, unless the burglar encounters one particular indicator: *noise*. Any kind of noise creates uncertainty and noise is the burglar's main concern and fear. In most instances, however, once the burglar has made a successful entry he or she will usually complete a burglary, unless the burglar is discovered by someone on the premises. Another factor that strongly affects the outcome of burglary attempts and the general level of burglary is police activity.

TOOLS OF THE BURGLAR

Some burglars prefer not carrying any tools, since, if caught, there is little evidence they planned to break and enter. But some will not consider a job without a good assortment of not only various sizes and shapes of screwdrivers, but glass cutters, wire coat hangers (easily shaped as the job requires), gear pullers, lock pullers, a hacksaw, a crowbar (especially helpful where considerable force may be needed), lock pickers, auto jacks, etc. And some burglars carry no tools at all. They may simply move from house to house or from apartment door to apartment door, trying for one that has accidentally been left open. When they find one, they ring the doorbell and have some sort of ready question to ask if someone does come to the door. If no one answers after repeated rings, then they are in and out in a flash, grabbing and taking anything that seems to have value. They are familiar with the most favored hiding places and are adept at locating a cache of jewelry and cash in a hurry.

When planning a job, the burglar sometimes phones to find out if anyone is at home. If nobody answers the doorbell, the thief takes out a finely honed screwdriver, slips it between the lock and door jam, and gives it a flip. The door almost always opens.

Good burglars must know their lights. A bathroom light on, according to this burglar, is hardest to figure. "You can't tell if anyone is home, especially if the bathroom door is left open so the light shines out. And timers that turn lights on and off give me fits."

FENCING STOLEN GOODS

There are a variety of routes through which the burglar disposes of the stolen goods. They range from the "square john" man on the street—who purchases, say, a color-television receiver at an abnormally low price, perhaps suspecting that the item has been stolen but reluctant to ask any questions—to the professional fence who "contracts for" large quantities of stolen goods to supplement a legitimate business or, in some cases, to operate a business entirely based on the sale of stolen goods.

For many burglars, a trustworthy fence is the key to a successful burglary. The fence provides the burglar with an outlet for stolen goods that, for various reasons, may not be desirable to dispose of through pawnshops or on the street. Furthermore, by being able to sell stolen goods immediately, the professional thief avoids the pitfalls of the novice who is very often caught with the items in his possession because he has no place to merchandise them quickly.

The exact relationship between the fence and the burglar is variable. The drug addict, desperate for a fix, may sell the goods at an extraordinarily low price, whereas the more highly skilled burglar, well-trained in the art of burglary, will often use more than one fence as an outlet for the goods, both to increase the bargaining position and to avoid the danger of losing an entire haul in the event that any single fence is caught.

TIME VERSUS THE BURGLAR

It would be a mistake to think that everything is in favor of the burglar, although the homeowner or apartment dweller is often most cooperative. Time is one of the burglar's greatest enemies. Even though ransacking a place that quite obviously has no occupants, the burglar does not know when they will return, or when the burglary may be observed by a neighbor. Some burglars allow themselves not more than a few minutes for each job. If, within the time limit they have set for themselves, they are unable to find anything of value or that they regard of value, they leave. No burglar will remain on a job indefinitely, unless some knowledge indicates that the occupants will not return for a long time.

CONFRONTING THE BURGLAR

What should you do if you surprise a burglar at work in your home or apartment? The first thing to do is not to panic. Most burglars do not carry guns, but in a tight situation they can make use of some heavy tools they may have in their possession as weapons. It may not be easy to remember as many physical characteristics as possible, but try to do so. Make a mental record of height, weight, possible nationality, complexion, color of eyes. Were the eyes normal, alert, or droopy? Any visible scars, marks or tattoos? Approximate age? Wearing a hat? Color of hair and way it is cut? Beard, mustache, sideburns? Shirt, necktie, jacket, or coat? Weapon, if any? Right- or left-handed? Kind and color of trousers? Shoes?

Yes, it would take an extraordinary individual to note and remember all of these items, and you may consider yourself cool, calm, and collected if you manage to recall half of them. But, we repeat, don't be a hero, and unless you are regularly accustomed to strenuous physical effort, don't try it at this moment. Don't use it as an opportunity to prove your superior power. If you have the burglar cornered, he will fight, knowing he could lose his life or spend many years in prison. All you are fighting for is property, and that's replaceable. And if you feel like screaming, don't. You will give the burglar no alternative but to silence you. And, as soon as it is safe to do so, call the police. If you don't remember the number to call, just dial the operator.

The best thing to do if you find a burglar in your home is to get out fast . . . if you can.

HOW NOT TO COOPERATE WITH THE BURGLAR

If you want to learn just how secure your home is, try breaking into it. People are sometimes forced to do this when they forget their keys or misplace them. Invariably they find a way of getting in, even though it may involve breaking a window, or borrowing a ladder and climbing in through a second-story window. But no matter how it is done, the point is that it can be done, and if the homeowner can do it, so can the burglar, and usually with less effort. It's all in knowing how, and know-how is the burglar's stock in trade.

There are many things anyone can do to make a house or apartment more secure. There is no such thing as a home that is absolutely impregnable; but that shouldn't be your goal anyway. What you should try for is not to attract the burglar's attention, but if having done so, to encourage the burglar to move on elsewhere.

1. When you are planning a trip, don't tell the burglar you are on vacation. Don't leave obvious signs of absence (Fig. 2-1). Stop

When leaving on a trip, be sure to stop deliveries that can advertise your absence.

Fig. 2-1.

your milk delivery, if you have it. Call your newspaper office and tell them you do not want delivery of the paper. Don't schedule it for the exact day you are to leave, but starting several days earlier. The reason for this is to make sure the newspaper will follow your instructions. It is entirely possible for someone to take your message and to forget or ignore it. Also, have the renewal date of delivery set for three or four days after your return, but not on the day you come back. Again, this is designed to give you a margin of safety in the event you cannot return on the day you expect.

Make arrangements with a cooperative neighbor to pick up your mail every day, or ask the post office to hold your mail or to forward it. Ask your neighbor also to remove any newspapers, circulars, or packages, and deliveries left at your doorstep. The newspaper may not be your regular paper, but may be instead a "free sample" or one that was left at your door in error. Don't just pick any neighbor, but one you can trust and preferably one who is obligated to you in some way, someone for whom you have done a number of favors.

2. For a burglar, the ideal arrangement is for you to leave your home and to leave it for a long enough time to provide the opportunity to ransack it without interruption. Nearly all homes are left vacant at one time or another. The time period may be short and you may be away only for shopping. But even if you plan to be away for only 15 minutes or so, at least lock your doors and windows (Fig. 2-2). Turn on a radio or a tv set, possibly with the sound up a bit louder than usual (Fig. 2-3).

Even if you may be away for only a few hours, lock your windows and doors.

Fig. 2-2.

3. You may have been advised to turn on a light and to leave it turned on while you are away. That's nonsense. Yes, do turn on a light, but remember that a strong light burning in your living room at three o'clock in the morning is a dead giveaway that you aren't home. Instead, buy several timers. These are inexpensive, simple devices that will turn lights on and off for you. Use one for a transistor radio set so that it will turn the set on during the hours that people habitually listen (Fig. 2-4). Don't set it for the wee hours of the morning! The station may not be on, and if it is, the sound (in the quiet of the night) will not only annoy your neighbors but will alert burglars that your home is ready for picking.

If you do use timers—and you should—put one lamp in the bathroom with the door kept open. Put the other in the kitchen.

Turn on a radio and increase the volume level. At night, turn on a radio and at least one light.

Fig. 2-3.

And use the third in the bedroom. Do not set the timers so they all go on and off at the same time. Arrange them so that as one turns on, the others turn off.

You will find an interesting variety of timers, but one of the more inexpensive types is the one shown in Fig. 2-4. It plugs into an outlet, while the controlled device, a lamp or radio, is plugged into the timer. Some timers of this kind are equipped with a lamp cord so they can be positioned some distance away from the outlet.

Another timer is shown in Fig. 2-5. The timer can be mounted directly into a switch box, replacing the existing switch. Since it is a permanent installation, it is much more convenient than the timer shown in Fig. 2-4. But it is also more expensive. The unit can be programmed to control lights or appliances to go on and off at various times, day or night.

Fig. 2-6 shows another programmable timer. One of its features is that it can automatically vary its on/off time, giving a home more of a lived-in look.

Some timers, such as the one shown in Fig. 2-7, are controlled by the amount of light in a room. Unlike other timers, no time setting is required. As soon as the room becomes dark the timer turns on. While the automatic feature is a convenience, the controlled light remains on all night, not a customary condition. It does have the advantage that once installed it requires no further attention.

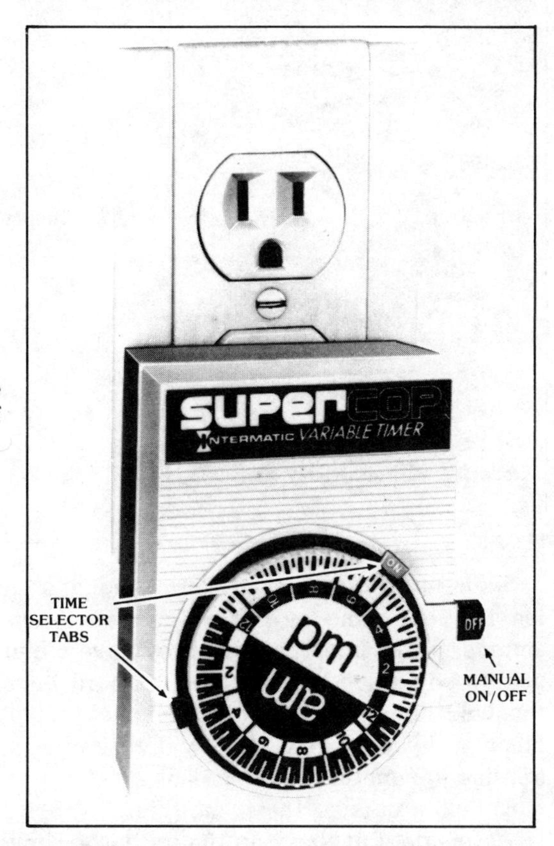

Automatic 24-hour timer can turn lights on and off during the day or night.

Fig. 2-4.

4. If you have a cassette recorder/player, make a recording of a dog barking. Do not record the barking continuously, but intermittently. A tape deck with automatic reverse is best for this purpose. The tape recording should consist of barking followed by a five or ten minute quiet interval. Make sure the barking sound isn't loud enough to annoy your neighbors, but strong enough so it can be heard by someone at your front door. You can also connect the tape player to a timer if you want a fairly long interval between periods of barking.

Solid state programmable timer fits into an existing switch box. (Courtesy Dynascan Corp.)

Fig. 2-5.

5. Do not pull down all your window shades whether you are leaving your home for a short time or are going on a trip. Keep the shades in the same position they have when you are home. How many people live in homes in which all the shades are pulled all the way down for twenty-four hours? The idea is to give your home a lived-in look that will discourage burglars. And this applies to venetian blinds, also.

6. Lights outside the home (Fig. 2-8) are just as necessary as those inside. Put these on timers but arrange for different on/off times. Lights in back of the house are just as important, if not more so, than front lights since the rear of a home tends to be darker due to the absence of street lighting.

7. Chapter 1 discussed the advisability of forming a protective group. If you have a private home, it would be well to start such a group with your neighbors on both sides of you (Fig. 2-9). Both sides are indicated here since the more friendly eyes you have watching your property, the more protection you have.

Another reason is that it is less likely that both of your neighbors will be away from their homes at the same time. Good

Fig. 2-6.

Programmable lamp timer controls up to 300-watt incandescent light and will repeat up to 48 selections. (Courtesy Dynascan Corp.)

neighbors help increase mutual security. If you must be away from your home overnight, keep your neighbors informed.

Also ask your neighbors to check your home regularly if you plan to take a trip. Write your itinerary, indicating where you are going, how long you will be away, plus a telephone number where you can be reached, if that is possible.

Encourage your neighbors to make use of your garbage pail. Spotters for burglars sometimes work for refuse collectors. Also ask your neighbors to be on the lookout for and to remove promptly any handbills, circulars, or penny savers that may be

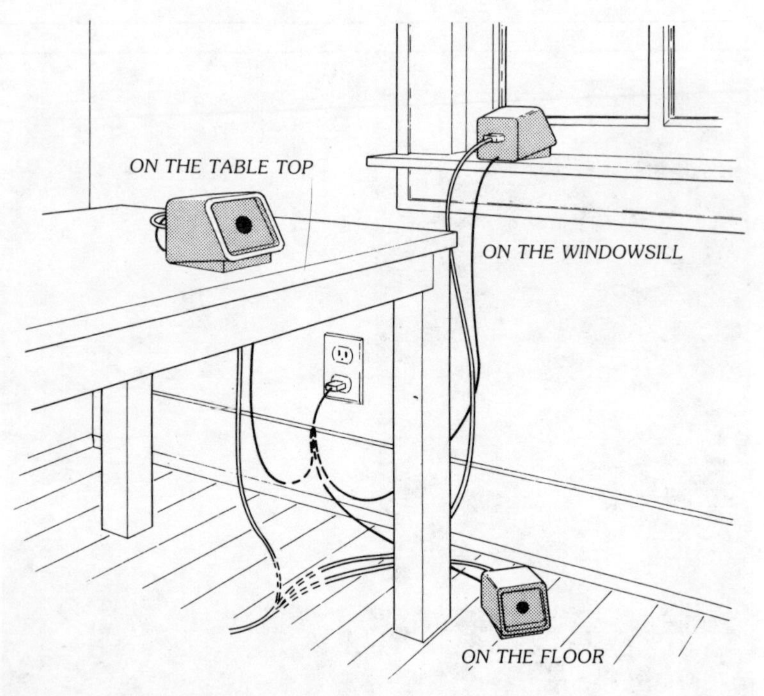

Fig. 2-7.

Lamp controlling device operates by amount of light in a room, turns on and off automatically. (Courtesy Cable Electric Products, Inc.)

Fig. 2-8.

Outdoor lights illuminate places of possible concealment. (Courtesy Cable Electric Products, Inc.)

Fig. 2-9.

Lack of concern for your neighbor's property makes any burglar's work easier.

left on your lawn, thrown on your porch, or placed outside your door.

8. Before you go on a trip be sure to close and lock all windows and doors, including the garage door. This sounds so basic you may wonder why it is included, but in the last-minute excitement of packing and departure, a window or screen may be overlooked. It pays to take the time to double-lock doors and double check everything.

Make a "security" check list just to make certain you do not forget anything. Prepare this check list several days before you leave and do some thinking about it between the time you first prepare it and the time you say goodbye. Then, just before you leave check off each item on the list to make sure you have followed your own instructions. Take the list along with you on your trip. There is nothing more guaranteed to ruin a vacation than to wonder whether you closed and locked the kitchen window, or whether you remembered to set the lighting timers.

Be sure to leave your house key with a neighbor. This can be a reciprocal agreement. Ask your neighbor to check your house occasionally to see if everything is all right. If possible, give your neighbor an emergency phone number where you can be reached. If your neighbor agrees to your "check my house" request, ask that the check be done in the evening as well as in the daytime. Most home burglaries take place in the daytime, but the nighttime act of turning lights on and off tends to keep burglars away. Also, ask your neighbor *not* to make the checks at the same time each day but to vary them. The checks should not be routine, but haphazard.

Notify your local police department. Tell them when you are leaving and when you plan to return. Let them know whom you have authorized to enter your home while you are gone. The police will make every effort to cooperate with you by keeping your home under periodic surveillance. If you must do this by telephoning your local police station, do so, but it is better to make a personal visit. This will not only make your request more emphatic, but will give the police an opportunity to learn who you are. Instead of just being a voice, you will be a person.

9. Most window latches aren't worth the room they occupy. All a burglar needs to do is to cut away a single pane of glass and he can then reach in and open the lock. Often, he does not even need to cut the glass. He just puts some tape on the pane and then smashes it with a rock wrapped in a cloth.

Window locks are described in Chapter 3, but you can make an inexpensive lock by drilling a hole through the top part of the window sash when both top and bottom windows are closed (Fig. 2-10). The hole should go through the top of the bottom sash and

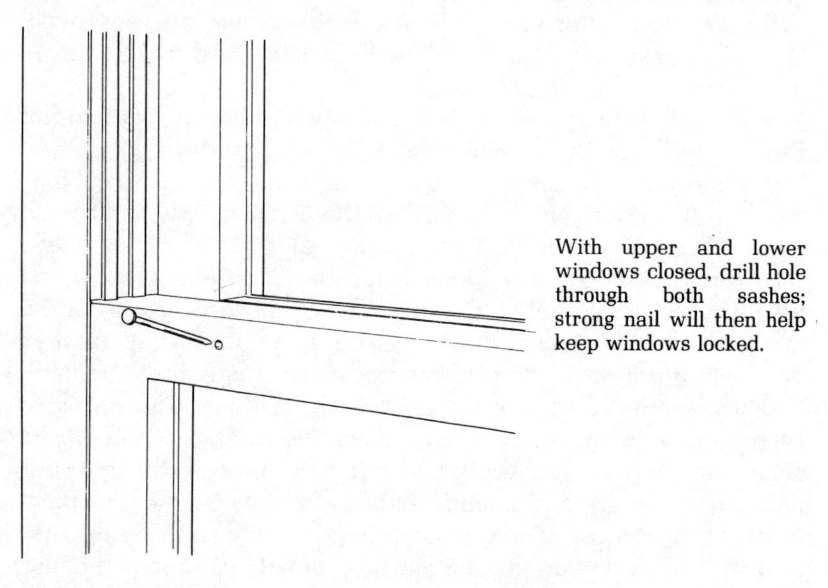

Fig. 2-10.

the bottom of the top sash. The hole must be large enough to accommodate a nail with the hole drilled at a downward angle to keep the nail from falling out accidentally. If you wish, you can drill the hole so that the nail head is covered by a curtain or drapery.

Be sure to use the thickest and strongest nail you can get, cutting it down to adequate size with a hacksaw, if necessary. A sturdy eyebolt is even better.

However, this arrangement still does not keep a thief from breaking a pane of glass, reaching in, and removing the pin. A window lock, combined with the nail arrangement, is better. You can also drill for a nail on both sides of the sash, at the extreme left and extreme right.

10. Don't tell strangers anything over the phone about your personal affairs (Fig. 2-11). If you are sociable, like people, and

Don't supply information about your home, property, or in-home time schedule to strangers making phone sales or surveys.

Fig. 2-11.

mingle with others, do so; but do not confide any details of your home or any of your going-away plans to anyone, except those neighbors who are going to watch your home for you. Actually, the fewer the number of people who know about your daily routine or a possible trip, the safer your house or apartment will be. There are some who must know—the police, your neighbor, and your employer, if you have one. And do not discuss your vacation plans on the phone. You can be overheard and you never know who is listening in on your line.

Your local newspaper may have a social column, listing the activities of people in your community. Do *not* contact them just to get your name in the paper. There will be enough time for that when you return.

Be careful to whom you send picture postcards while you are away. The best method is to time the mailing of the postcards so they are delivered just as you return home.

11. Don't keep stocks, bonds, or other valuable papers (Fig. 2-12) at home. Do not keep spare cash at home. Deposit your spare cash in your bank and put your papers and valuables in a safe-deposit box in your local bank. Do not put cash in your safe-deposit box. Put it in your savings account.

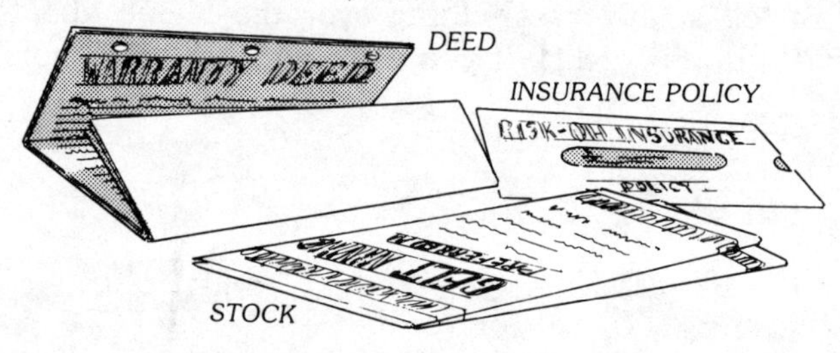

Fig. 2-12.

Keep important papers in a bank safe-deposit box, not at home.

12. Use an engraving tool (an electric pencil) to mark all your equipment (Fig. 2-13) with your name, or your Social Security number, or both. This may not add much to the appearance of the equipment, but it will make it much more difficult for the burglar to fence, plus the fact that the fence may either not accept such equipment, or may offer a substantially lower price.

13. You may pride yourself on having an excellent memory but don't depend on it as a substitute for a list (Fig. 2-14). Keep sales slips for anything you buy of value together with a list of your valuables in a safe-deposit box. Be sure to include serial and model numbers. This will make it easier for you to get your claim for burglary loss approved by your insurance company.

14. If you have a dog and expect to be away for a short time, possibly just a few hours, don't take him with you. He is excellent house and apartment protection, since he is a noise producer. Some dogs are trained specifically for security purposes, but any dog, provided he likes to bark, is useful.

15. If you have a house, equip it with outside weather-resistant

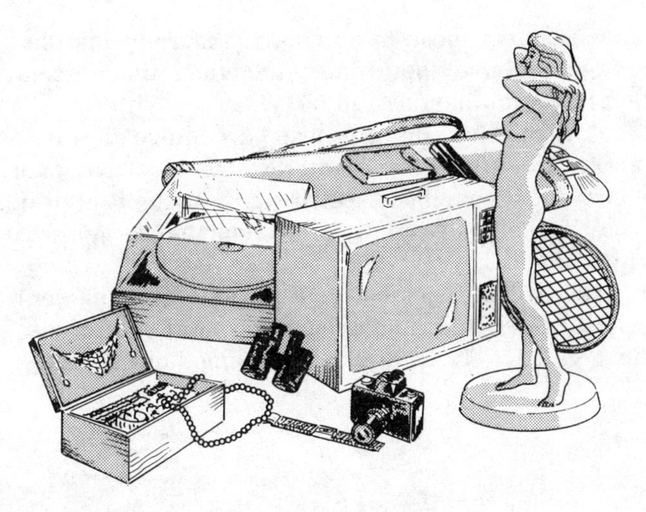

Fig. 2-13.

Mark your name or social security number on valuable items with an electric pencil.

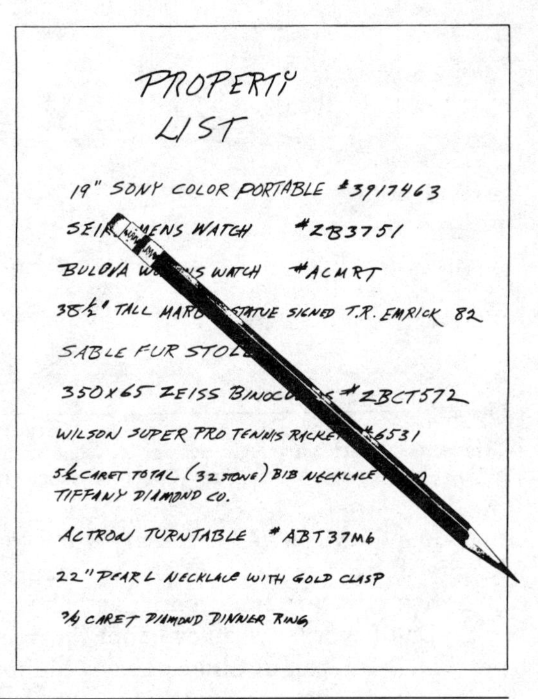

PROPERTY
LIST

19" SONY COLOR PORTABLE #3917463

SEIK[illegible]ENS WATCH #ZB3751

BULOVA W[illegible]NS WATCH #ACMRT

38½" TALL MAR[illegible]STATUE SIGNED T.R. EMRICK 82

SABLE FUR STOL[illegible]

350X65 ZEISS BINOC[illegible] #ZBCT572

WILSON SUPER PRO TENNIS RACKE[illegible]6531

5k CARET TOTAL (32 STONE) BIB NECKLACE [illegible]
TIFFANY DIAMOND CO.

ACTRON TURNTABLE #ABT37M6

22" PEARL NECKLACE WITH GOLD CLASP

¾ CARET DIAMOND DINNER RING

List your valuables and keep the list somewhere outside your home, preferably in a rented safe-deposit box.

Fig. 2-14.

floodlamps. Connect these to automatic timers inside the house, and arrange to have them come on when the automatically controlled lights in the house go off.

16. Most burglaries, the greatest percentage by far, are by direct entry by the burglar right through the front door. If a burglar can enter your home via the front door it is preferable rather than forcing a window, an action that is more likely to attract attention.

There are various types of systems for protecting your front door, consisting of lights, an alarm, or a combination of the two as shown in Fig. 2-15. Such devices are sometimes equipped with

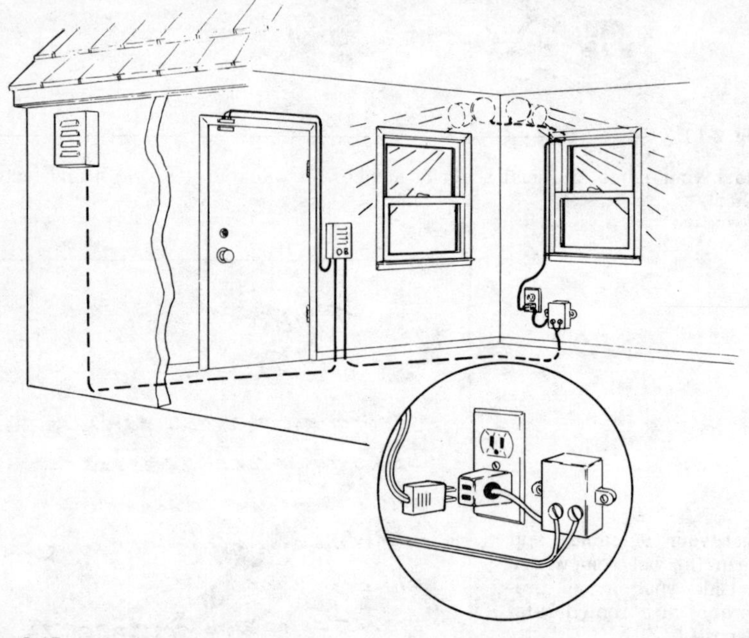

Fig. 2-15.

Typical combined alarm/light system — when door is opened, system goes into action after predetermined time. (Courtesy Hydrometals, Inc.)

time-delay devices and a defeat button that give you enough time to enter the house and shut off the alarm system. The system must be activated every time you leave home.

17. Don't ever leave a key under the door mat or under a flower pot (Fig. 2-16), or any other obvious hiding place. If, for practical reasons, you must do so at least select a place that will be difficult

Fig. 2-16.

Do not hide house key under door mat, flower pot, or other such obvious place.

for the burglar to find. For apartment dwellers good hiding places are practically impossible; plus the fact that, sooner or later, someone will see you hide or retrieve the key.

18. Always keep in mind that burglary is quite often a crime of opportunity and whether it is successful or not depends on how vulnerable you are.

19. Never, under any circumstances, leave your credit card (or cards) at home.

20. Burglars respect properly installed safes and locks, provided these are the right types. A burglar can look at a lock and learn from it just how security-conscious you are. There is probably no lock a determined burglar cannot open, but burglars must work against time. If they abandon an attempt at entry, it isn't because they cannot overcome the lock; simply that it could take much too much time to do so.

21. If you move into a new home or a new apartment, either have the lock tumblers reset, or better yet, install new locks (Fig. 2-17). Do this even if you have been supplied with keys to an existing lock. You do not know who else may have duplicates of those keys.

Fig. 2-17.

When you move into new house or apartment, replace all locks; or have all tumblers of existing locks reset.

22. Make sure you have a chain lock on the front door like the ones in Fig. 2-18. As shown in Fig. 2-18, the device is being used

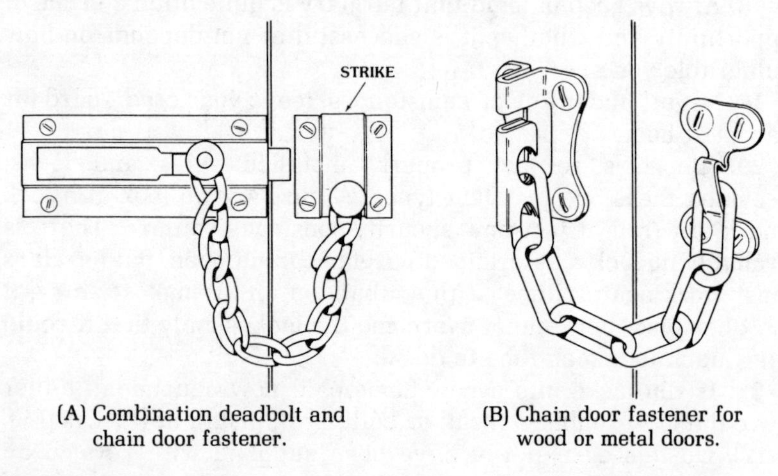

(A) Combination deadbolt and chain door fastener.

(B) Chain door fastener for wood or metal doors.

Fig. 2-18.

Two kinds of chain guards. (Courtesy National Manufacturing Co.)

as a chain guard only, but the strike can be turned 90° so that the unit can be used as a deadbolt. The chain is made of case-hardened steel. The chain door fastener in Fig. 2-15B is for wood or metal doors and uses a case-hardened steel welded chain that can be hooked into any desired link to adjust the door opening for viewing or ventilation. A determined burglar can make short work of such a lock simply by kicking in the door; however, the lock is helpful if you want to talk to people at your front door but do not want to open that door to them.

23. Many people have a habit of combining their house keys and car keys on the same key ring or in the same key holder (Fig. 2-19).

Keep your house and car keys on separate key rings.

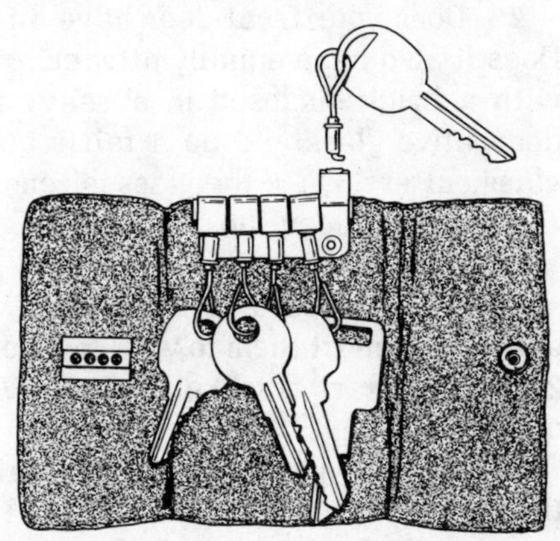

Fig. 2-19.

Keep your car keys and house keys on separate rings. If you must leave your car key with a parking-lot attendant, leave only the car key. It is very easy for anyone with a mind for doing so to duplicate your house key. And your home address can be obtained from your license-plate number.

24. If you use an electronic telephone-answering device, always put a message on the tape that you expect to return momentarily, even if you expect to be away for several days. Never supply strangers with personal information. A common trick is to use the "survey" approach. "We are making a survey for (and here the caller will mention a well-known company) and would like to know . . . ." The caller may or may not represent the company named. Do not supply financial information, data about your possessions (number of cars, tv sets, cameras, typewriters, etc.). If such callers are persistent, and they may well be, just hang up.

25. Do not admit strangers to your home. They may seem respectable, pleasant, charming, and well dressed. But so are a lot of burglars. They may be giving your home a pre-examination to determine if it is worth burglarizing. Always keep the chain lock on. This will limit the amount of door opening but will still let you communicate with someone at your door. You can ask for identification but don't put much reliance in it. IDs can be easily faked. At least obtain a telephone number so you can call for verification, but while you are doing this, the person at your door should remain outside, with your door shut and locked. Do not rely on the chain lock.

26. Does your front door have an attractive large glass panel? Does it also have equally attractive side glass panels? A burglar with a brick enclosed in a heavy rag can make short work of decorative glass and do it fairly quietly. Many are experienced glass cutters. Once the glass is removed (Fig. 2-20), a burglar can reach in and open the door lock unless the lock requires a key on both sides of the door. Use a metal grillework to cover the glass.

27. If any street lights on which you depend are broken or inoperative report them to your town or city authorities at once (Fig. 2-21). Burglars dislike lights. Why improve their working conditions?

28. Be aware and beware of strangers, particularly if you see them loitering near your home or the home of a neighbor. If you are at all suspicious, telephone your neighbor and your local police department so you can both keep an eye on the activities (Fig. 2-22). Burglars often case or examine homes they regard as likely prospects for their activities. The burglar may be well-dressed, appear conservative and drive a well-kept, expensive looking car. He may well look more like a very successful businessman or an executive than a burglar.

29. When not in use, keep your patio doors locked, preferably from the inside. A strip of wood on the inside bottom track will help prevent the door from being opened from the outside.

You can also use a Charley Bar. This is a metal bar, pivoted at one end. While it is more convenient than using a strip of wood on the bottom track it serves the same purpose. A better arrangement is to use both the wood strip and the Charley Bar. If you do use a wood strip make sure it is thick enough so it will not bend when pressure is applied to open the patio door. A cut-down broom handle will do if you can trim it to fit in the space. Make

Glass doors and side panels can easily be broken. Use metal grilles over all glass areas; also use lock that is key operated from inside and outside. (Courtesy Noblit Brothers & Co.)

Fig. 2-20.

sure it is shorter by about 1 inch than the space it is to occupy, so that you can remove it easily when you want to open the patio doors. You can also use a length of ½-inch iron pipe (Fig. 2-23). Another method is to use a patio door lock, but make sure it is the type that locks.

30. When you leave your house, close your garage doors (Fig. 2-24). Open garage doors and an empty garage are a positive indication of an empty house.

31. If you've been painting or repairing around your home put all ladders away when you are finished with them (Fig. 2-25). Put them inside and store them so that they cannot be easily removed. As an alternative to the nuisance of carrying ladders in and out of the house, lock them with a steel chain and a sturdy lock to some

Fig. 2-21.

Report broken street lights if they have not been replaced within a few days.

Fig. 2-22.

Report strange or unusual activities in your neighborhood to the police.

sort of strong fastener. This will also help keep your ladders from "walking away."

32. If you must do some work in your basement, in your attic, in your garage, or in any area that is away from the house proper,

Use strong wood or metal bar or pole in channel at bottom of your patio door.

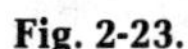
Fig. 2-23.

Fig. 2-24.

Keep garage doors closed except when garage is being used; never leave your home with garage doors open.

consider your house as unoccupied and act accordingly (Fig. 2-26). Lock your front door, and any other doors, and keep your windows locked. If you have a dog that can act and behave as a watchdog, don't have him keep you company. Let him stay on guard duty where he belongs.

33. Do not leave any valuable articles outside your home. This includes bicycles, lawnmowers, tools, and children's toys (Fig. 2-27). Burglary can take place right outside your home, as well as inside. If you do not plan to use your car, keep it inside the garage. If you do intend using it, keep the car doors locked and the car windows closed. Cars can be and are being stolen from the driveway. Even if you leave your car just for a few minutes to go

Fig. 2-25.

Never leave ladders outside overnight. If you must, chain them securely.

Fig. 2-26.

When working in the attic, basement, or garage, keep front and rear entrances locked.

Fig. 2-27.

Put away folding chairs, lawn furniture, toys, bicycles, lawn hose, etc., before retiring for the night, or when not being used.

into your house, act toward your car as though you were going to leave it for hours.

34. Try not to get into a regular "I'm leaving the house" routine. If possible, try to leave at different times and try to return at different times. Give your house the appearance of being lived in constantly. The greater the traffic flow in and out of your home or apartment, the less the chances for a burglary. If your family has a number of members, try not to leave together and try not to return together. An exodus, en masse, from your home can inspire only one interpretation.

35. If you live in an apartment and the mail is delivered after you leave for work in the morning (as it probably is) try to arrange with a neighbor to remove your mail as early in the day as possible. All a burglar has to do is check the community mailbox in an apartment house to learn which apartments may be vacant and which may be occupied. If you are a single woman, never put your full name on the mailbox (Fig. 2-28). Use your first initial and be sure to omit Mrs., Miss, or Ms. Physically, an unarmed burglar will feel he is more than a match for any female. There is nothing wrong with the "fem lib" spirit. Women are fully entitled to all male privileges and prerogatives, but identifying yourself is a

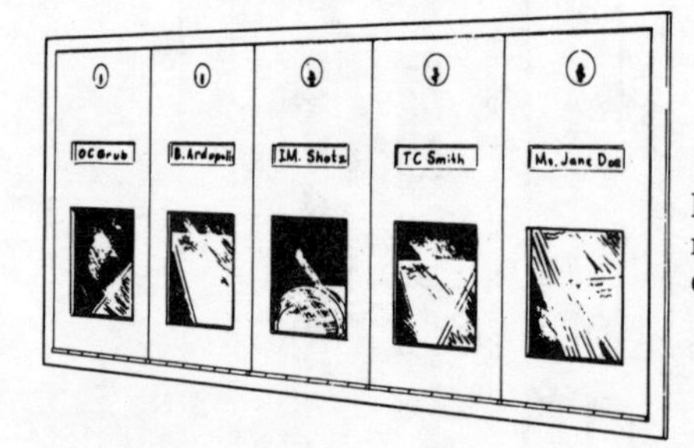

Instead of Mrs., Miss, or Ms. and a first name, a woman should use an initial only on the mail box.

Fig. 2-28.

poor tradeoff for which you get nothing in return, except possible trouble.

36. If you live in an apartment house, form a mutual-protection society with the neighbors on your floor. In many apartment houses the occupants tend to be withdrawn and unsociable, but for your mutual protection you should set up a patrol system so that corridors are watched. You will also find it helpful to post a sign in the corridor that it is being watched and that all the neighbors on that floor are highly security-conscious. Report any strangers on your floor to the superintendent at once. Organize a "buddy" system when you go shopping or use the inside laundry in your apartment house. And remember, never enter an elevator with an arm full of packages together with a stranger. Try to arrange to do your shopping and your laundry with a neighbor, and preferably with a group.

37. Two locks are always better than one (Fig. 2-29). Put supplementary locks on any window or door that gets constant usage. Be especially sure to lock any windows that open onto or are near fire escapes. In apartment houses, bathroom windows near fire escapes are sometimes left unlocked and are often used as a means of entry by a burglar. Even if the window is several floors above ground, lock it. If you have doors or windows facing an outside balcony, and that balcony is several stories above ground, lock them. A competent burglar doesn't always use an elevator or stairway, but can move from one floor to another by climbing up outside balconies.

38. If you must have maid service in your home or apartment, be sure to check references and do so thoroughly and carefully (Fig. 2-30). Never put temptation in anyone's way. Keep your liquor locked up, and do not keep cash and/or jewelry in an easily

Put additional lock keyed from inside only on doors leading to rear garden, terrace, or balcony.

Fig. 2-29.

If you hire in-home help, always check references carefully.

Fig. 2-30.

opened drawer. The advantage of having a maid working in your home is that it is occupied during her working time. The disadvantage is that the maid is a stranger. It is better to hire someone recommended by a friend or neighbor and who has worked for them for a number of years. Even if the hourly rate is higher, it is safer. If possible, do not have the maid report on a regular schedule, but try to stagger her working hours from week to week.

39. If you have a private house, do not put your name on your mailbox or place an identification plate on or at the sides of your front door. Unwanted solicitors often look for name plates and use the name to establish an immediate confidential relationship. Beware of solicitors who claim they have been recommended by your neighbors. Your name posted on or near your house makes your telephone number readily available to anyone.

If you have a mailbox mounted on the front of your house immediately adjacent to the front door, try to have this replaced by a slot-type arrangement that is cut into the door so mail is fed directly into the house. The advantage is that if mail is delivered at a time when no one is home the presence of a stuffed mailbox will not act as an advertisement that your home is empty.

40. If you live in an apartment house the main door should have a buzzer type lock (Fig. 2-31). If it does not, get together with the other tenants in the house to demand that one be installed. If the apartment house front door is so equipped, do not answer the buzzer automatically. Do not open the door unless you know who is requesting admittance.

If an apartment house, always make sure main front door locks after you use it.

Fig. 2-31.

41. If you must be away from your home for a number of days, arrange to have your lawn mowed (Fig. 2-32). An overgrown lawn is a grand advertisement that your home is vacant. Using a car, burglars sometimes cruise a neighborhood looking for a house that is easy pickings. In the wintertime, arrange to have your walk (or walks) and driveway cleared of snow. The important point is to give your house a lived-in look. Use services that are advertised in the local paper and phone directory or that have been recommended by friends or neighbors.

42. Your house may have a rear basement door or some other door that is seldom used. Such doors can be easy points of entry

Fig. 2-32.

Arrange for yardwork on regular schedule while you are away.

for a burglar. An inexpensive way to secure such a door is to put a hasp and lock on it as shown in Fig. 2-33.

A sturdy hasp and lock on seldom-used doors adds security.

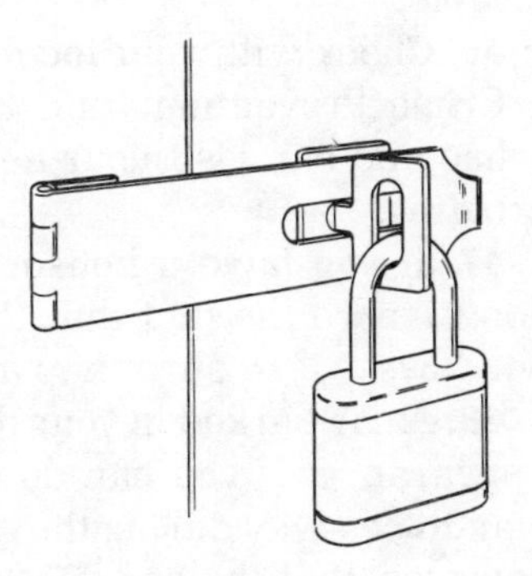

Fig. 2-33.

43. Type or write the phone numbers of your local police and fire departments on a small slip of paper and paste this paper on your telephone or on something close to it. This will save you precious minutes during an emergency.

Find out if your area is serviced by a 911 emergency number. This is much easier to remember and to dial than the usual 7-digit code. Not all police departments have the same 911 equipment, but in some cases dialing 911 tells police personnel the location and phone number of someone calling for help and displays this information on a monitor screen, somewhat like a television set. In the stress of a situation such as a burglary, the victim can dial 911 and then hang up without supplying a name or address. With such a system the police do not need to ask questions but have the basic information they need to provide help. The best way to find out is to visit or phone your local police department and learn what information they have when someone dials 911.

44. If you have a house, make sure that shrubs and other landscaping do not obscure or cover front or rear entrances. Entrances should always be as visible as possible and not offer a hiding place. If you do have dark areas, see what you can do to provide illumination.

45. People with children to protect often use baby sitters. Why not try a house sitter, someone who would occupy your home for the evening you expect to be away. Consider also using a house sitter when planning a trip of one or more days. Make sure you know something about the house sitter. A neighbor would be preferable.

46. Check with your local police department to see if they have a Crime Prevention Unit. Such units are staffed by experts in crime and can visit your home and make suggestions for burglar proofing.

47. If you have a house with a basement and a second floor, don't ignore the fact that these home areas also have doors and windows. Make sure they are securely locked.

48. A car parked in your driveway will give your home a lived-in appearance. If you can do so, keep a car in the driveway if you plan to be away during the day or on an extended trip. Make sure, however, that the car doors are locked and that all car windows are completely closed. You can get window security devices in an auto parts store that make window opening with a coat hanger impossible.

49. A covering of snow that has no footsteps or which has not been shoveled for several days is a good indication of your absence. If possible, arrange to have snow removed even if you are away just for the day. This tactic is even more essential if you plan to be away for a longer time.

50. Take a photograph of your valuable possessions (Fig. 2-34) before leaving for an extended time period. You can arrange

Fig. 2-34.

Itemizing valuables after break-in could be extremely difficult, if not impossible; if items are fragile, a picture inventory of them is advised. (Courtesy Eastman Kodak Co.)

small items on a table or other flat surface to be photographed as a group. On the back of the photo write the model number and serial number (if such numbers are available), date of purchase, and price. Take a picture of your house from all sides, and include pictures of any outbuildings. Photograph the interiors of all rooms so that all items in each photo can be easily identified.

Large objects may require that you take pictures from several angles for complete coverage. Once this in-house picture file is complete, with information such as date of purchase, etc., on the backs of the prints, store the negatives and at least one set of prints in a secure place off the premises.

If you have a video camera, you can make a film record of all your possessions and use it for proving losses because of fire or burglary. Pan each room with the camera, moving the camera slowly from side to side, followed by a closeup shot of each item of value in the room. You should include a voiced description of each article, date of purchase, model and serial numbers, distinguishing characteristics, and a general description. Also include any special features which will help in their identification.

A recorded video tape of this kind will last indefinitely. A good method is to keep the film in a bank vault with an identification label on the tape indicating its function and the date on which the video tape was made.

Since your possessions will change over a period of time, it is a good idea to make new video tapes, possibly once a year. Before doing so, play back the old tape as a reminder of what you have done. If you have disposed of any items that appeared in a previous tape, indicate that on the new tape.

A taped record of this kind is generally accepted by insurance companies as proof of claim, particularly when substantiated by written records such as sales slips. Such written supporting data should be kept together with the video tape so that you have a unified loss package.

You can also use either photographs or video tape as evidence for a tax loss. A before and after picture on video tape is far better supporting data than an explanation.

If you have a break-in, and after the police have arrived and made their investigation, take a photo or make a video tape record of the premises. If there has been any vandalism, take pictures of that as well.

These 50 examples have not exhausted all that are possible

because your home may require special security measures. You can add to these to take care of your own requirements. And, as you can see, not all security measures mean you must spend money.

Finally, cooperate with the police and any other law enforcement agencies. Become involved. Just because a burglary happens next door to you is no reason to give a sigh of relief. Instead, look on it as a warning. A burglar likes to pick off his victims one-at-a-time. Why let the burglar have his or her own way?

BEDROOM SECURITY

In a sense, when you retire for the night, your house or apartment becomes vacant for you then occupy just one room. If you do use just one bedroom, consider the advisability of equipping that bedroom door with a 1-inch deadbolt lock, a thumb turn type that fits on the inside of the door. No key is required and there is no access to the lock from the outside.

This does not mean you must lock yourself in every night, but there is no reason why you should not if the added security makes you more comfortable. No key is required for such a lock and it can be opened and closed easily and quickly. The advantage is that if you hear a burglar in some other room closing and locking your bedroom door will give you some time to call the police. It is always better to avoid a confrontation with someone who has forced entry into your home.

This arrangement works well if you are living alone and there are no other bedrooms in the home that are occupied. It also assumes that you have a telephone in your bedroom.

YOUR HOME SAFE

Is it debatable whether having a safe in the home is desirable or not. Some police departments advise it, others counsel against it. If the safe can be hidden in some way so its presence isn't obvious, and if it is well constructed and installed, and if it is fire retardant, it can have advantages. A free-wheeling type floor safe, however, is an advertisement that you have objects of value. Such a safe may be noticed by visitors or repair persons and it can then become a lure.

THE SECURITY CLOSET

If you have objects of value that are only used from time to time, such as camera equipment, a microscope or telescope, or skis, you should consider the possibility of setting up a security closet.

This closet should be equipped with a keyed type of 1-inch deadbolt lock. Since closet doors are outward opening types with readily accessible hinge pins you will need to replace these with hinges that are concealed. Keep the key to your security closet on your key ring together with your house keys.

Many homes have barely enough closets and it is a rare home that can spare a closet just for security. In that case, you may have a closet that is mainly used for storage. If so, consider the possibility of making it do double duty as a security closet.

WINDOWS

According to the FBI, burglars, vandals, and other intruders break into over 2 million American homes every year. A determined invader gets in through doors, windows—any and all openings in your home. And the rising crime rate shows that it takes more than just locks on your doors to stop these break-in artists. This means that your windows must also be protected, and this applies equally to upstairs windows, basement casement windows, and garage windows, as well as to ground or main-floor windows. There are any number of devices that can be fastened to windows that will make the burglar's job a more difficult one. Windows usually come equipped with a center latch, but this latch is inadequate for a number of reasons. It only works when the windows are completely shut and so it is ineffective during the warmer months. Also, when the interior window frame is painted, the latch is often painted over and so it gradually freezes in the open position. Further, it can be opened by any burglar who wants to go to the slight trouble of breaking a window or cutting it open.

There are a number of window-locking supplements, generally available in hardware stores. One is a screw type lock (Fig. 2-35) while another is a thumb-operated type that works by depressing it at one end. Both of these lock the upper and lower window together. The advantage is that these devices permit the window to be opened any amount and still supply a locking feature. They have the disadvantage, however, that they can cause the window

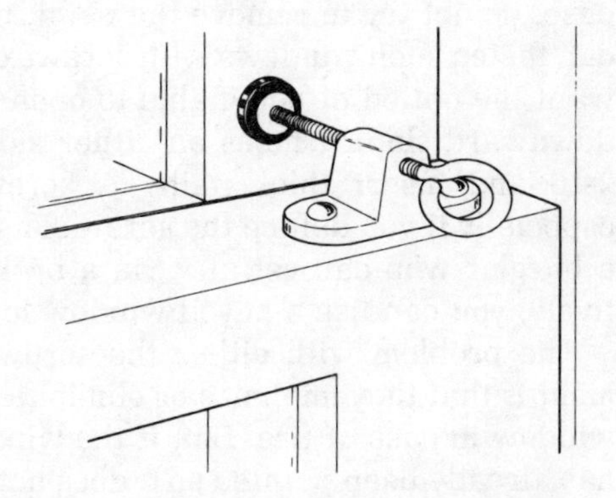
Screw-type window lock.

Fig. 2-35.

to warp and can be opened by any burglar who can remove a windowpane.

Another window protective device includes the window jammer (Fig. 2-36). This wedge-shaped device is usually mounted

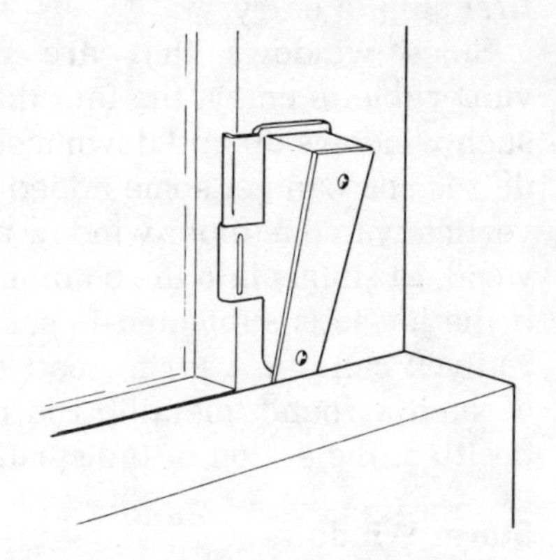
Window jammer is simple, with no moving parts; however, it must be repositioned each time window is used.

Fig. 2-36.

on the upper right-hand side of a window a few inches above the center bar. In the closed or locked position, the unit will block and stop any attempt to open either the top or bottom window. However, it can also be defeated by any burglar getting past the window glass.

In a house, much more so than in an apartment, there may be some windows that are never used. For such windows, it is advisable to use screws to keep them shut. Commonly, such windows have a sash that has been painted over and so it is difficult, if not impossible, to open or close them. If that is the case, do not try to remove the paint, but don't depend on it. You can fasten such windows with screws on both sides, but if you still want the option of being able to open the window, drill a pair of downward sloping holes on either side as shown earlier in Fig. 2-10 and insert thick nails. A screw arrangement is better, especially if you deface the screw slots. A nail can be removed by a burglar who can get at it via a broken pane of glass. Alternatively, you can use a keyed window lock.

The problem with either the screw or window lock arrangement is that they minimize or eliminate the possibility of using the window in case of fire. But, if the window is seldom used, or if it has already been painted so it does not move up or down easily, it is unlikely it would be used in case of fire. In any event, the window can be smashed open. If you do use a keyed window lock, keep the key handy somewhere near the window so that you can always use it, if necessary. However, be certain it is out of the reach of someone from outside who removes a pane of glass as a first step to entry.

Since windows that are used regularly make the home vulnerable to entry, the fact that it is convenient for you to slide such windows up and down means it is also easy for a burglar to do so. You can get some added security by putting a wooden rod vertically in the upper window channel. The rod should be of hard wood, fit snugly into the channel, and will remain in position even if the house is subjected to some vibration. Instead of a wooden rod you can use a spring-loaded window-curtain rod. The length of such a round, metallic rod is easily altered and when put in position, the action of the spring helps keep it in place.

Storm Windows

The problem with ordinary windows is that they interpose just a single sheet of glass between the would-be intruder and the victim. Storm windows not only provide added insulation, but can help form a more effective barrier to entry.

Most storm windows aren't lock equipped, but are made to permit the window to stop and hold at various points. Unfortunately, storm windows lose much of their burglar-deterrent effectiveness during hot weather when a screen is put in position in place of the glass pane. A screen is not much of a block since it can be cut and removed as easily as glass. However, during cold weather months, the pane of glass of the storm window is helpful for keeping intruders out.

Casement Windows

There are various types of basement casement windows. Some are made of wood only, others are metal. Metal types are preferable provided that they are adequately secured. Some casement windows are made to swing outward and upward, using hinges across the top. Others are operated by turning a small handle that can open the window.

Don't rely on the fact that such windows are generally small, since a determined burglar of the right size will be able to squeeze through. Some burglars form teams consisting of an adult and a child, because the child is able to get through very small openings.

You can cover small windows with a wire mesh so that breaking the glass does not permit access, but still allows you to open and close the windows. How effective the wire mesh is depends on how thoroughly you fasten it and the kind of mesh you use. As a further precaution, equip the window with a barrel bolt. You can also use a strong metal grille across the window on the outside. This will not interfere with the movement of the window if it is the type that opens inward. Since the grille will be fastened with screws, damage the screw heads so that they are unable to be turned with a screwdriver.

Street Level Windows

Some houses, whether apartment houses or private homes, sometimes have windows that are at street level. These windows put the resident at a considerable disadvantage since any passerby can look in and gauge the possibilities of a break in.

For such windows, use a keyed sash lock plus pins on both sides. Another arrangement is to use a bar set consisting of a number of metal bars mounted in a frame that can be hinged to the side of the window. The bar set is kept locked by using a padlock. In high crime areas, it might be necessary to use wire mesh, making sure it is securely nailed into place. While the bar set and wire mesh screen are unsightly, a choice must sometimes be made between security and appearances.

Louvered Windows or Doors

The problem with a louvered window or door is that it supplies some security when all the glass slats are closed, but when opened, individual slats can be removed. For added protection, the inside of the door or window should have a decorative grille covering over the entire length.

A louvered door is usually equipped with a simple latching lock. The purpose of the lock is to keep the door closed, but not locked. In that case, consult with your local locksmith for replacing the latch with a more secure type of lock.

Louvered doors are generally used as entrance doors to a closed porch. Once the burglar gets past the louvered door he can work on the main entrance door knowing he can no longer be seen. Thus, a louvered door equipped with a latch lock and opening into a closed porch supplies the burglar with ideal working conditions. The louvered door, then, is your main defense against entry and not the front door.

If the porch has louvered windows, it is advisable to keep the louvres in their closed position except when the porch is being used. If you want the advantages of ventilation, the best arrangement for louvered windows is to install interior metal grilles.

WINDOW LOCKS

The window latch can be a keyed type, but even here the screws that hold the latch in place don't provide much security.

Aluminum Window Lock

The key-operated window lock shown in Fig. 2-37 mounts at either top corner of a movable aluminum window and fastens it to the window jamb. When activated, it locks the movable window in either the closed, or partially open, position. A tempered locking

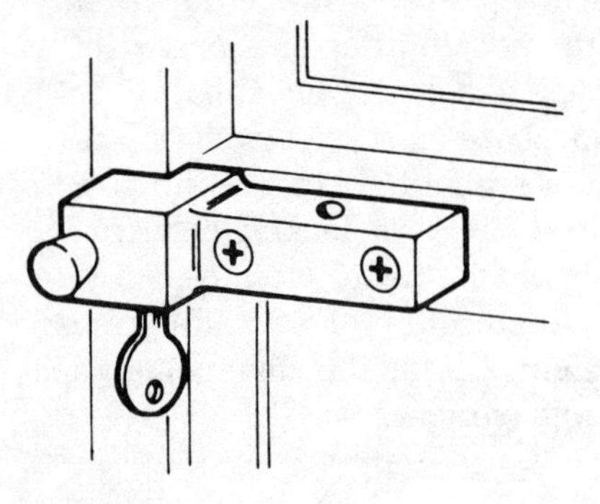

Key-operated lock for aluminum windows. (Courtesy Taylor Lock Co.)

Fig. 2-37.

bar installs through the lock and into the jamb. A five-wafer cylinder holds the locking bar in place.

Wooden Window Lock

The keyed lock shown in Fig. 2-38 is designed for windows made of wood. The lock mounts on the top corner of an inner window

Key-operated lock for wood windows. (Courtesy Taylor Lock Co.)

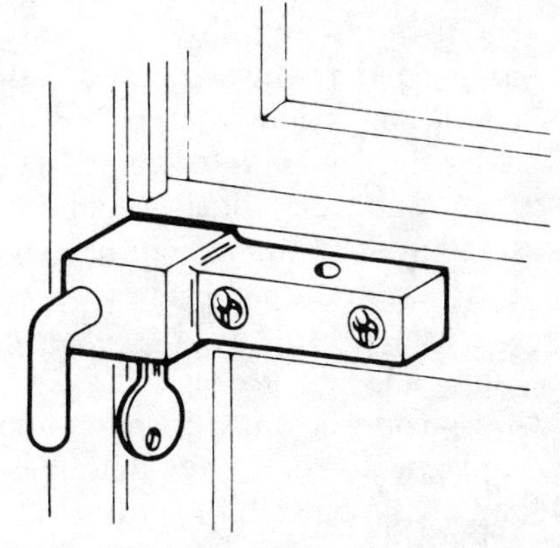

Fig. 2-38.

and, when activated, prevents both the inner and outer windows from being opened. Designed as a supplementary lock for sash locks or hook latches, this window lock can be installed to lock the windows in a fully closed, or partially open, position.

THE TELEPHONE AS A SECURITY DEVICE

The telephone is often overlooked as a security device, but it is as important as any lock or light. If you have a choice between a rotary dial telephone for your bedroom and a touch tone, select the touch tone. If at any time you need to make an emergency call when you suspect someone is in your house, the touch tone phone is much quieter and quicker, and you may be able to make a call for help without attracting attention.

You should always:

1. Keep emergency numbers next to your telephone—police, fire, ambulance, and physician. If you have 911 emergency service in your area, make certain all members of your family are aware of it. Telephone the office of your local police to find out just what services the 911 number supplies. In some areas, all you

need to do is to dial 911 and hang up. In others, you must give your name, address, and indicate the nature of the emergency.

If the emergency is a fire, alert everyone in your house, get out and call for help from the home of a neighbor.

2. Keep an up-to-date list of all your credit card numbers near the phone as well as the numbers to call if your credit cards have been lost or stolen.

3. Have the number of your local drugstore or pharmacy nearby so that your doctor's office can call if a prescription is needed. Learn how ambulance service is supplied in your area. Learn what number or numbers to call if you need ambulance service and keep these numbers adjacent to your phone. Also keep the phone number of a friend or relative, someone you can depend on in an emergency, next to the phone. Don't depend on your memory for this. In an emergency, we tend to forget numbers we may use regularly.

4. Consider getting a Touch-a-matic telephone. With these you can program emergency numbers into the phone so that with the touch of one button, the call will be placed immediately. Numbers can be color coded for younger children.

5. Train your children to say that Mother (Father) can't come to the phone right now, instead of, "Nobody's home but me."

Nuisance Telephone Calls

An unwanted telephone call can be the beginning of a scam to involve you financially, or it can be an annoyance call or an obscene call. The best way to handle such calls—the technique recommended by telephone companies—is to say nothing and to hang up.

By hanging up your telephone you take away the basis, the reason that prompted the call in the first place. The moment you hear and become aware that remarks are being made, hang up.

If, after you answer the phone, all you hear is silence at the other end, hang up. People you know and people who are interested in contacting you always start talking immediately. Silence at the other end of the line indicates a crackpot call. Do not slam the telephone receiver into place since such an angry reaction will indicate to the person at the other end that they have managed to get through to you. Don't yell, don't speak, other than a single hello. Do not say "who is this?" Above all, do not repeat the question. The one word, "hello" is all that should ever be necessary.

If you find yourself the target of repeated annoyance calls, notify your local telephone company.

Wrong Numbers

Do not give your telephone number to someone who has called the wrong number. Instead, tell him he has dialed incorrectly or has the wrong number. If he is calling long distance, give your area code and just the first three digits of your number. In that way, the caller can get credit for his wrong call.

If You Are Disabled

Check with your local telephone company to find out which special services and equipment are available. If you have a hearing problem or you have impaired speech there are aids that amplify either the voice being received or the volume of the speaker. There is also equipment available that will convert sounds into sight signals so you can see when the phone is ringing.

For those with impaired vision one-number dialers permit automatic dialing of pre-programmed emergency or frequently called numbers. Touch-Tone service is also used by many visually impaired people to eliminate the need for finding dial openings.

Speaker phones and headsets that are available in very light-weight models provide hands-free use of almost any type of phone equipment.

THINGS YOU SHOULD NOT DO

While there are numerous things you should do to make your house or apartment more secure, there are also some things you should not do.

Do not ever leave your home, whether house or apartment, with the front door unlocked. Some people have the attitude, and it is completely wrong, that if a burglar is going to get in he is going to get in. That is not only nonsense, but is incorrect. Some burglars operate simply by trying doors. If the door is unlocked and if no one answers the front door bell, they are in like a flash. If the door is locked, they may move ahead to another house or apartment.

If you are a gun hobbyist, keep your guns locked up or at least do not display them on walls which can be seen from the street. Burglars are attracted to guns.

If you use one or more window fans make sure they cannot be removed easily. Such windows should be equipped with some method for locking. Window fans often have a pair of sliding panels, one on each side, and their purpose is to make installation as easy as possible. They may be easy to install, but they are also just as easy to remove.

If you live alone, try to keep that information to yourself as much as possible. Do not mention it to delivery people, appliance service technicians, or anyone else you do not know personally.

Never carry identification tags on your key ring or key holder that will identify you, your car, or where you live. If possible, avoid carrying your keys in your purse.

THE PROBLEM OF KEYS

If you live in a house rather than an apartment, it is likely you will have accumulated a collection of keys, so many in fact that carrying them all on a key ring or in a key case becomes a problem. Aside from the fact that a collection of keys can be heavy and a nuisance to carry, there is always the problem of finding the right key when you want to open a door.

A solution is to carry only those keys absolutely required, such as your car door and trunk keys, and your front door key. All the other keys can be left at home, but this raises the question of where and how.

One method is to put a label or tag on each key and then hang these keys on nails or hooks at some convenient location in the house. The labeling or tag enables you to identify each key quickly, but if it is convenient for you, so it is for a burglar. It may seem odd that a burglar should have any need for your keys if he has already made his entrance, but there is always the possibility he may have come in through a window or patio door. He would much rather leave by your front door since this method of exit is least likely to arouse suspicion from a watching neighbor. If your front door has a lock that is keyed on both sides, he will need a key to make his exit that way. Also, he may want your keys so as to make a return trip to your home at some other time. Burglars do make return visits.

Instead of putting labels and tags on the keys, just put a number on each. Make a list of these numbers and keep the list in some place in the home, possibly behind a row of books, or in a book or under a corner of a rug. This doesn't mean the burglar won't find it, but at least you will have forced him to search. Searching takes time and that is precisely what he doesn't have. Put the keys where you will be able to get at them but where the burglar will need to search to locate them.

Decoy Keys

Another method is to have a number of keys, none of which fit any locks in your house. These can be odd ones you have collected or which can be obtained from friends, neighbors, relatives, or a locksmith. Put labels and tags on these, clearly identifying them as front door, rear door, basement door, etc. Put your good keys elsewhere, in a more secure location and identified by labels or numbers. There is no question that the burglar will find the decoy set and will try them. But all this will take time, enough to make the burglar apprehensive and that is exactly what you want to do. You may not want a personal confrontation, but neither do most burglars. They prefer in and out, as quickly as possible, and with no fuss.

Your Front Door Keys and Shopping

If you do your grocery shopping on the way home from work, you should understand you are putting yourself in the best position for being mugged. Further, if you start looking for your keys while shifting packages you are again in a prime position to be attacked.

Try not to load yourself down with packages and arrange your shopping so you do not carry much at any one time. Use a shopping bag, preferably one made of cloth, hence much stronger than paper, and better because it has a handle. In this way you can easily drop the bag and have both hands ready for defense and at least one hand available for reaching for your keys. A heavy load, in paper bags, requiring the use of both hands, is an attractive setup for a mugger.

INSIDE YOUR APARTMENT OR HOUSE

The fact that you are inside your apartment or house does not mean you can let down your guard, that you should lose your awareness. There are two kinds of thieves who will call you to your front door. One is the person who claims to be collecting for a charity of some kind, or who mentions the name of one of your neighbors, obtained easily enough. There are all kinds of scams including those who claim to be collecting for a local fire or police benefit, or who carry Bibles and ask for donations to a local church, or who use the name of a well-known organization, such as the Boy Scouts, the Girl Scouts, etc.

In some instances the scam is conducted by an adult, with a child making the donation plea. A good answer is to ask for an envelope, informing the inquirer that you will mail your donation. No matter how persuasive the person at the door may be, your answer should be a firm no. Do not buy any home or apartment services from front door solicitors. You will be told all kinds of stories—you have been especially selected, the sale is part of an advertising plan, only you can take advantage of the very low price—whatever. Just ask that all advertising literature be mailed. Don't be misled by the fact that the solicitor may be a young girl or a pair of young girls, or what seems to be a husband and wife team.

After you get the literature and get a chance to read it, call your local Better Business Bureau and find out more about the product, or service, being promoted.

LOBBY STICKUP

Stickup men, burglars, and muggers often make use of a lobby for their crimes. The lobby is hidden from the eyes of neighborhood crime watchers. The victim, having entered the lobby has a feeling of security and is no longer alert. Further, a crime in the lobby means the perpetrators can use your keys, force you into your own apartment, and not only relieve you of cash and credit cards, but any portable objects of value. You are also much more subject to violence.

The fact that an apartment house has a doorman does not always supply as much security as you would think. When the doorman opens the door for you, it is possible for a pair of attackers to push you and the doorman into the lobby. A protection technique is for the doorman and the tenants to be alert to men hanging around the entrance or near it.

ELEVATOR CRIME

Elevators in apartment buildings and office buildings, as well, are dangerous locations. If you go into an elevator, it is better if you go in with a group. Watch as the floor-numbered buttons are depressed. If yours is the highest number, get off at a lower floor rather than remaining on the elevator with an unknown person. If you are the only person left on the elevator and a stranger enters at some floor before you reach yours, press the elevator button for a stop at a floor, or floors, below the one the stranger has chosen.

Groups of young criminals, possibly three or four, will sometimes get on an elevator with a prospective victim, but will select an exit that is one floor higher. The victim, lulled by this ploy, moves ahead to his (or her) apartment door, finds the door key, opens the door, and is suddenly surrounded by the young people who have raced downstairs.

WHAT YOU SHOULD DO BEFORE OR DURING AN EMERGENCY

The best thing to do about an emergency is to try to avoid it. Upon returning to your house or apartment, be alert to the possibility that someone may have forced entry while you were gone. Don't just rush into your home. Take a quick glance at your windows and your door to see if you can observe anything unusual or different. Some burglars leave doors and windows open, not as their calling cards, but as a means of quick exit should it become necessary. If you do see some evidence of entry (Fig. 2-39), stay out and immediately get to a telephone and call

Fig. 2-39.

If you've been robbed, disturb nothing; call the police immediately.

the police. They are experienced and have weapons and know how to use them if they are necessary. Further, you won't have to explain anything to the police should you succeed in capturing and overcoming the burglar (which is unlikely).

PROTECTING THE BEDROOM

If you are a light sleeper, if you are disturbed by the slightest noise, or if you fall asleep aware of the possibility that you may be burglarized, you can be sure of one thing. You aren't going to get a good night's sleep since all the conditions are against it.

Under these circumstances, the best thing to do is to install a lock on your door. Don't use a key-type lock, since, in case of fire, you may want to get out of the room in a hurry. Instead, install a barrel bolt. These are slide bolts, usually cylindrical, and can be screwed to the door and door jamb by wood screws. The strike is that part of the lock which receives the bolt. It can be a universal type, mounted into the door jamb, or a surface type that mounts on the jamb. The universal type is more difficult to install but it is stronger and is more push resistant than the surface strike. Barrel bolts come in various sizes but the smaller ones, usually about two inches, can be ripped off the door with one strong push. Buy a sturdy one that is about 5 inches long. And use it. It isn't intended as a decoration. Barrel bolts are designed with cutouts holding the handle of the bolt in position, so these bolts cannot be pushed back by an intruder on the other side of the door (Fig. 2-40).

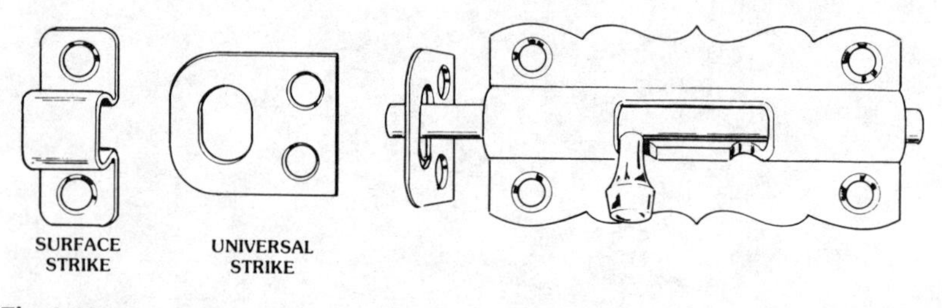

Fig. 2-40.

Barrel bolt can offer temporary security for bedroom door. (Courtesy Stanley Hardware, Div. of The Stanley Works)

Barrel bolts are intended for use with all-wooden doors. If your bedroom door has glass panes, it would be relatively easy for a burglar to remove the pane, slip his hand inside, and slide the bolt back. For a glass door, then, you will need some type of keyed lock. Be sure to mount the key somewhere near the door, but well out of the reach of a probing arm. Remember where the key is if you should need to leave the room in a hurry.

As an alternative to the barrel bolt, you can use a swinging hook and screw eye (Fig. 2-41). The swinging hook is fastened to

Swinging hook and screw eye does not offer as much protection as barrel bolt but is inexpensive. (Courtesy Stanley Hardware, Div. of The Stanley Works)

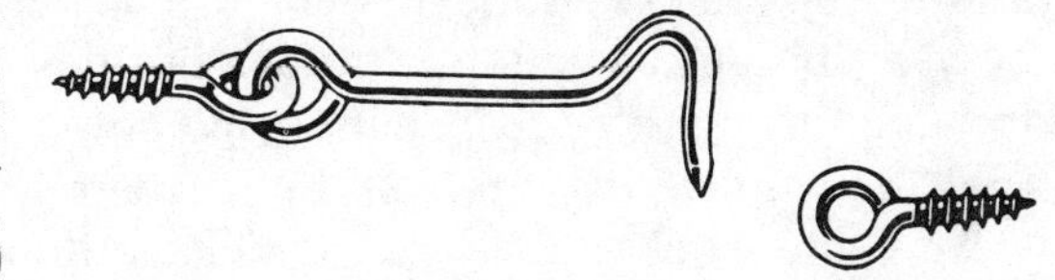

Fig. 2-41.

the door; the screw eye to the door frame. This is easier to mount than a barrel bolt, and costs less, but doesn't provide much protection. A burglar can slip a thin knife or shim between the edge of the door and the jamb and lift the swinging hook out of position. But if a swinging hook is what you prefer, get one that is fairly substantial. The small ones pull out much too easily.

If you do have a bedroom lock of some kind, get into the habit of making sure that the bedroom door is locked before you retire. There is no point in having a lock unless you use it. And if you have a lock, be sure to have a telephone in the bedroom. This will enable you to phone for help. Otherwise, you will be a prisoner in your own bedroom while the burglar ransacks the rest of the house.

You can also get a very economical, battery-operated door alarm for your bedroom door. Such door alarms are completely self-contained. They include a buzzer, a battery, and a switch that is closed when the bedroom door is opened. The entire unit is small and can be mounted directly on the door. The door-operated alarm—actually just a buzzer—is no substitute for a true alarm system. The sound it produces will be enough to awaken you, but it may not be loud enough to deter a burglar. It all depends on the burglar, his determination to rob you, and the kind of valuables he expects to find. It also depends on the kind of person you are. Many people prefer sleeping through a burglary, and would rather not have the shock of an encounter.

Even with a telephone and a barrel bolt on the door you still have a problem. Although you phone the police immediately, there is no assurance they will arrive before the persistent burglar gets through the door. A professional burglar, hearing you telephone the police, will make for the nearest exit—and quickly. He or she is smart enough to know that the risk factor has suddenly gone up considerably. Your thief, though, may be an amateur; may be high on drugs or alcohol; may be a mental case; may think you are alone, weak, helpless, frightened; or, may not think at all, but will be in a fit of unreasoning rage. The burglar may be strong enough to push through the door, not caring how much noise is made, or that the police may very well be on their way.

As protection, you should have something with which you can defend yourself. Keep a golf club handy. Or, a broom handle. Or, a length of round metal rod of the kind used for drapes. Or, a baseball bat. Whichever one of these you select, do two things: (1) Keep it somewhere near your bed; and (2) get accustomed to handling it. Do not keep a loaded gun in or near your night table unless you have a permit for the gun, unless you know how to use one, and unless a gun is part of your usual business activities.

Chapter 3

Locks

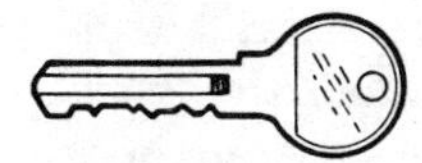

Doors are the most frequent points of entry for residential burglaries. There are a number of good reasons why a burglar will select a door as his entry (and his exit) area. It may be habit. We are more accustomed to going in and out of doors than through windows. Further, the entry through a window may not be convenient—that is, the window may be too far above ground. And the burglar knows that departure via the front door is least likely to attract unwanted attention. Finally, the burglar knows that if the robbery is successful, both arms may be loaded with your household goods. It is desirable to get out with as much as possible and so a door is the logical exit.

This means, then, that in a security setup in the home, the front door requires the most attention. This is particularly true in apartment houses, where windows may not be accessible to the burglar, leaving him no option. The burglar, of course, can trick you into opening the door, and so having lock security isn't always enough. You must be on guard against being pressured or talked into opening your front door. There is a good rule to follow, though. When in doubt, keep your door locked. It may later prove to be an embarrassment, but this is the lesser of two evils.

DAMAGE TO YOUR HOME

About two-thirds of the burglaries reported in one study indicated there was no property damage; where damage did occur, the value of the amount of damage was usually fairly low.

In some instances, a burglar may do more damage to a door in making entry than inside the house or apartment. A burglar may resort to violence if the haul does not seem to be enough.

People who have very modest apartments or homes sometimes feel they are safe and secure, since their possessions are of little value. But for the most part, burglaries do involve modest-cost items. Burglars will go for typewriters, tv sets, costume jewelry, hi-fi components, but if these aren't available, will steal anything that isn't fastened down. The reason for this is that the burglary is being done by a nonprofessional. The professional burglar selects a victim and makes sure that the robbery is worth the risk. In some instances there may be an opportunity to visit the house or apartment allowing the burglar to mentally assess all of those items having a high-market value. The nonprofessional, however, breaks in and hopes for the best. And if that home doesn't offer the best, then he takes whatever he can.

In a survey made of a particular area, the following information was obtained:

- In three cases, all residential, the premises were occupied at the time of the burglary or burglary attempt.
- Of the eight residential burglaries that occurred while the premises were unoccupied, the premises were left vacant for one hour or less in three cases.
- In two of four attempted burglaries, the potential intruder failed to gain entry because of effective preventive measures in operation.
- In only one residential burglary was there a dog on the premises at the time of the burglary.
- In two of five nonresidential burglarized sites that had alarm systems, the burglars managed to bypass the alarm.
- In six out of nine cases in which someone saw or heard the burglar, the observer was a neighbor.
- In half of fourteen cases in which the victim had an idea who committed the offense, the suspected individual was a young neighborhood resident.

The trouble with these statements is that they are generalizations. They do not hold for all parts of the nation, and may vary from one neighborhood to the next. They do have some value in that they may supply some idea of the best defense to take.

THE VICTIMS

Here are a few generalizations about victims of burglaries.

- Victims are more likely than nonvictims to be aware of a general crime problem in their neighborhood. Once someone becomes a victim he is apt to become a concerned citizen.
- Victims of burglary are more likely to be victims of other crimes than are nonvictims of burglary. Just as there is such a thing as being accident-prone, so too is there such a thing as being "victim-prone." Some victims, after a burglary, become very security-conscious, but as the incident fades into the past, they resume their former careless habits and so suffer another burglary, and then another.
- Nonvictims are more likely to take simple precautions against burglary than are victims. This may sound like a contradiction, but the nonvictims are nonvictims for the very reason that they do take precautions.
- Burglarized structures are more likely to be on corner lots.

THE COST OF PROTECTION

Whatever devices you install to protect yourself and your family against theft and fire have a payoff in the form of peace of mind knowing that you are saving both lives and property. Some insurance companies will also give you a discount on your insurance premiums if you install protective units. The following list indicates the discounts made available by one nationally known insurance company, with discounts up to 15 % on a homeowner's policy base premium. If you qualify for such a discount, you should contact your insurance company or agent.

If You Have:	You Save:
1. Smoke detectors on all floors, a fire extinguisher, and dead-bolt locks on all exterior doors	5%
2. Local burglar alarm that is installed on all windows and doors on the first floor	5%
3. Both 1 and 2	10%
4. Burglar alarm that is installed on all windows and doors on the first floor, and reports to the police station or a central station	10%

5. Fire alarm that reports to the fire station or a central station 10%
6. Both 1 and 4 .... 15%
7. Both 1 and 5 .... 15%
8. Both 4 and 5 .... 15%

THE BURGLAR'S TOOLS AND METHODS

The concept of a burglar carrying a bag of sophisticated tools is one that has been fostered by tv. Many burglars do not carry tools for two good reasons. First, there is no reason for him to burden himself. Often enough he will find the front door of a house either open or so poorly protected that getting in presents no problem. Sometimes just a well-placed kick or two, correctly done, will force many doors to yield. While this does produce noise, something no burglar enjoys making, a few kicks are unlikely to arouse suspicion.

The second reason is that no burglar wants to carry incriminating evidence with him. Further, he wants his hands free to be able to carry heavy objects from your home. Your silverware, television sets, hi/fi system and bric-a-brac require the use of both hands. However, burglars may have available an assortment of useful small tools including the "cheater" or "shove knife," a *shim* made of light metal or plastic, screwdrivers, glass cutters, lock pullers, vise-grip pliers for twisting doorknobs and wrecking the interior mechanisms, lock picks, and fishhooks for use as lock picks.

Still another tool is the jimmy, or pry bar. A jimmy can be used to separate a door and its door jamb so that the bolt of the lock is withdrawn from its strike, a rectangular plate that fits into the door jamb and receives the bolt of the lock. The jimmy permits the use of considerable force and isn't a noise producer. Further, it can be hook equipped so the burglar can easily carry and conceal it under his clothing.

There are three other tool techniques that are used. An expert burglar can cut away a pane of glass in a matter of seconds, and, with the help of rubber suction cups, pull the glass outward quietly. He can then reach in and undo most locks (not all) that bar his entry. He may be equipped with a small, sharp saw for cutting thin wood panels, or, if circumstances permit, may even use a hammer.

SUBTERFUGE TECHNIQUES

Many burglaries take place in the daytime with the burglars resorting to some sort of subterfuge for getting into a house or apartment. In one case, burglars bought tickets to a highly popular show and mailed them to the residents of an apartment with the compliments of the theater manager. The thieves watched the "lucky" couple as they left their home and knew exactly how much time they had for their burglary. In another example, burglars parked a large crate marked "television receiver" right in front of an apartment door. One of the burglars crept into the crate and then, unobserved, was able to saw his way in through the front door.

The Baby Trick

In one method, a man and a woman, or two women, use a baby as a means of getting into the houses or apartments of elderly women. They explain that the baby is hungry and simply ask for permission to warm the baby's milk bottle. After gaining admission, one of the two thieves goes into the kitchen and, of course, is accompanied by the occupant. The other thief is then free to ransack the place.

Telephone Call Trick

Some small residence hotels and rooming houses have a single telephone for the use of occupants, generally located on a lower floor. A thief will get the names of the residents, and then telephone them from an outside corner telephone. His associate, knowing the full details of the call, is at the victim's door as soon as he leaves and forces his way in. Quite often the victim, thinking he will return in a few minutes, leaves the door unlocked. The victim is kept on the phone long enough to give the second thief enough time to sweep through the rooms to pick up anything of value.

The Inspectors

All a thief needs to do is to put on some kind of uniform and wear a cap with the word "Inspector" prominently displayed on it. He can then claim to be a wiring inspector, a sanitation inspector, a house-safety inspector, a boiler inspector, a water inspector, or any other kind of inspector. He then suggests that he be allowed to make an inspection, as required annually by the laws of the city or town. If he is denied admission, he may mumble something about the possibility of a fine for noninspection. Once inside the house he usually looks around for any item of value he can pick up.

ELEMENTS OF A LOCK SYSTEM

It is estimated there are 3,000,000 burglaries per year for all of the states, but it is thought that an equal number go unreported. Every day $75,000 is stolen from American homes, with one burglary committed every 15 seconds, 24 hours a day. Every home is a potential target. Many homes, even if equipped with locks, are left unlocked or have entry areas that are weakly protected, if at all. In many instances, the locks that are used are so poorly installed or are so basically wanting in design that a burglar can overcome them in a matter of seconds without using tools.

By itself, no lock, no matter how well made, is burglar proof. The area of entry, whether door or window, must be considered as a unit. A strong lock with a poor door is still inadequate, for the combination can be no better than its weakest element. These elements include the lock, the door, the door frame, the strike, and any accessories that may be used to reinforce any one or all of these elements.

Finally, even the best combination will not thwart a determined burglar who has enough time plus some assurance, often supplied by the homeowner, that he will not be disturbed. But a well-thought-out combination will usually encourage a burglar to move on to a less well-defended home. There are so many available he has no need to waste his time, skills, and efforts.

Basic Rule for Security

A basic rule for houses and apartments, and one often overlooked is—lock up. Residents of houses and apartments cooperate with burglars to a surprising extent for in almost 50% of reported burglaries a door or window was left open, or a key was put in an obvious hiding place.

A lack of security can be due to poor habits. A good habit to get into is to lock up before you leave, even if you plan to be away just a short time. That short time can easily stretch into a time that is long enough for a burglar to get in and out; particularly if you conveniently and cooperatively supply an easy entrance.

The Purpose of a Lock

The purpose of a lock is simple. Its basic function is to keep the burglar out, and, insofar as it is capable of doing this, it will give you a sense of security. That's a psychological bonus, but whether you really earn it or not depends entirely on the kind of locks you use and whether you give them a chance to do their job—that is, if you keep them locked when they should be. A lock, the proper lock, is your first line of defense against burglars.

There is no lock a determined burglar cannot open. But to most burglars there are three essential requirements to the success of his act: time, quiet, and darkness. It is not that a burglar cannot pick a particular lock you may have selected, simply that the burglar cannot afford or is unwilling to trade the time against the possible risk of detection. And there is no such thing as a pick-proof lock. If you see one advertised as such, then the manufacturer is either deceiving himself, deceiving you, or both. A reputable manufacturer will advertise its locks as pick-resistant. But that leaves the manufacturer with an escape as wide as a barn door: How pick-resistant is pick-resistant?

TYPES OF LOCKS

There are basically two types of locks: *active* and *passive.* Very few active locks are made and only on order for highly valuable, highly specialized commercial protection. An active lock is one that fights back. It may contain a gun that will fire at any individual not authorized to open it. It may fire a gas, or spray a chemical mist. Some oldtime locks even had spring-loaded steel blades, ready to fly out at any intruder. The modern active lock (for the most part, but with a few exceptions) sets off an alarm or sends a signal to some externally located security office. Some locks are now being offered for home use that turn on the lights in all or a part of the house if tampered with; or that ring a bell, buzzer, or some other kind of alarm; or that may even connect with a security office or the police department.

Under the category of passive locks there are hundreds of variations, including mortise cylinder, rim cylinder, deadbolt cylinder, profile cylinder, drawer locks, padlocks, alarm locks, cam locks, key-in-knob locks, thumb operated locks, switch locks, furniture locks, window locks, patio door locks, and a number of slide bolt types.

Another, somewhat simpler, way of categorizing locks is by the way in which they are made part of the door. Thus, there are two basic types. The mortise lock (Fig. 3-1) and the rim lock (Fig. 3-2).

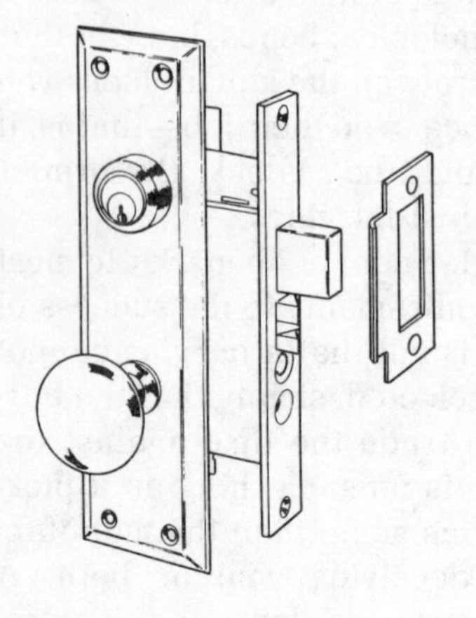

Mortise lock. Cylinder is keyed and operates the deadbolt. (Courtesy Noblit Brothers & Co.)

Fig. 3-1.

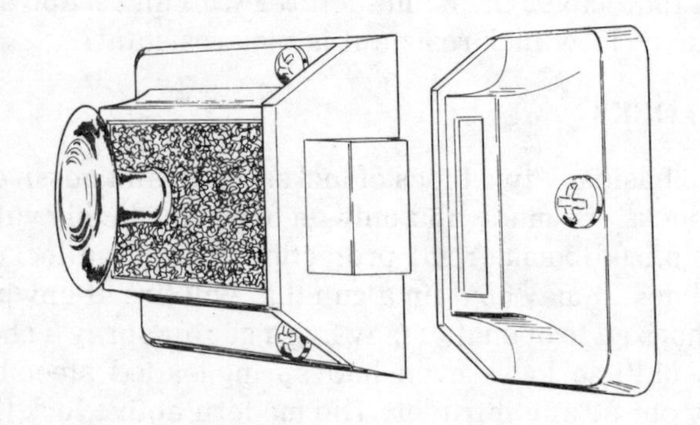

Fig. 3-2.

Rim lock with square deadbolt, if equipped with cylinder, can be key operated from outside, thumb operated from inside. (Courtesy Noblit Brothers & Co.)

The mortise lock requires that a hole, or a groove or a slot—some opening—be cut in and through the door. The lock actually becomes part of the door. Similarly, the strike for a mortise lock, the part of the lock that receives the bolt, is mortised into the door jamb. The mortise for the strike is cut sufficiently deep so that the strike forms a continuous surface with the door jamb.

The rim lock, unlike the mortise lock, is fastened directly on the door, and is held in place with screws or bolts. Instead of a strike, the rim lock uses a *keeper*, although it performs the same function as a strike. The keeper receives the bolt (or bolts) of the rim lock and is not mortised into the door jamb.

There are advantages and disadvantages to both the mortise and rim locks. As the rim lock is easier to mount, cutting into the door or jamb may or may not be needed. Further, the keeper of the rim lock is generally much stronger and much more substantial than the stroke of the mortise lock. However, the rim lock is quite unsightly.

The use of one of these locks, such as the mortise-type, does not eliminate the use of the other. Thus, a front door may have a mortise-type lock using a single or double cylinder plus a rim-type lock, having just a single cylinder mounted on the inside.

PARTS OF A LOCK

Both locks, mortise and rim, are somewhat similar in the kind of parts they use. A mortise lock is not necessarily superior to a rim lock, or vice versa. Both are essential, if a home is to be well lock equipped.

The Lock Bolt

The *bolt* of a lock is the moving portion. The *strike* is the part that receives the bolt and is fitted on the door jamb. The *door jamb* is the wooden frame that faces the lock.

The lock bolt can be a single or double length of metal. It has various shapes and can be round, rectangular, or semicircular. It can be rather thin, or quite thick and substantial. The lock bolt can be reinforced by a section of special heat-treated, cutting-resistant steel through its center.

The Strike

The strike can be nothing more than a rectangular cutout in the door jamb, but the usual strike is like either of those in Fig. 3-3, which shows a rectangular section of metal with (A) one or (B) two cutouts. The edge of the strike may be curved to keep the bolt of the lock from hitting the jamb.

The strike is usually held in place with flat-head wood screws, having a length of 1 or 2 inches. A screw length of 3 inches is preferable for outside doors. The best arrangement is one in which the screws are long enough to reach in and penetrate the stud for at least 1 inch.

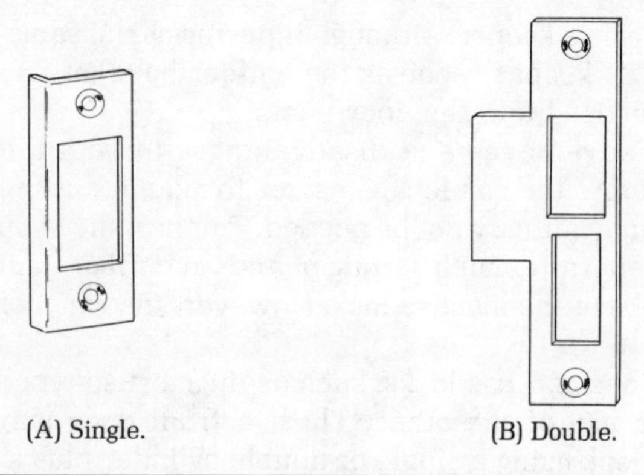
(A) Single.
(B) Double.

Fig. 3-3.
Two kinds of strike plates. (Courtesy Taylor Lock Co.)

The strike is often a weak point of the lock system. Several well delivered kicks can shatter most strike arrangements. A reinforced strike, as shown in Fig. 3-4, consists of a metal housing positioned behind the faceplate of the strike. A reinforced strike is superior to the flat-plate strike.

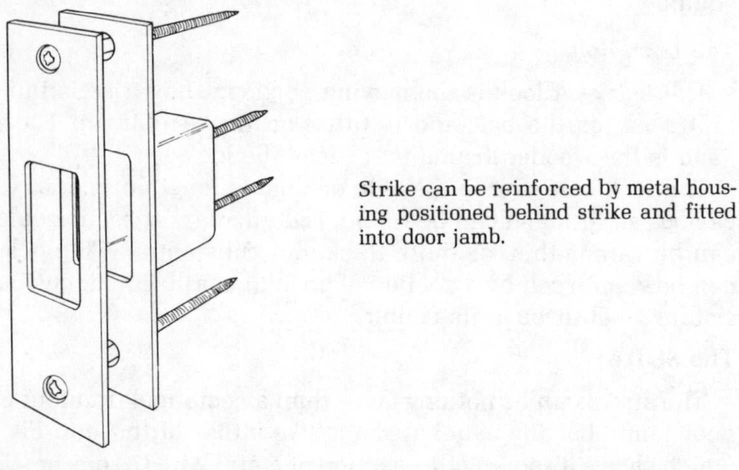
Strike can be reinforced by metal housing positioned behind strike and fitted into door jamb.

Fig. 3-4.

The Keeper

A keeper is generally supplied as part of a lock set when you buy it and it can be made to receive a horizontal bolt or vertical bolt. Fig. 3-5 shows a vertical rim lock with its keeper. Keying the lock moves the bolt vertically into the keeper which is held in place on the door jamb by flat-head wood screws.

Rim lock with vertical bolt has double cylinders and so is suitable for doors with glass panels. (Courtesy Taylor Lock Co.)

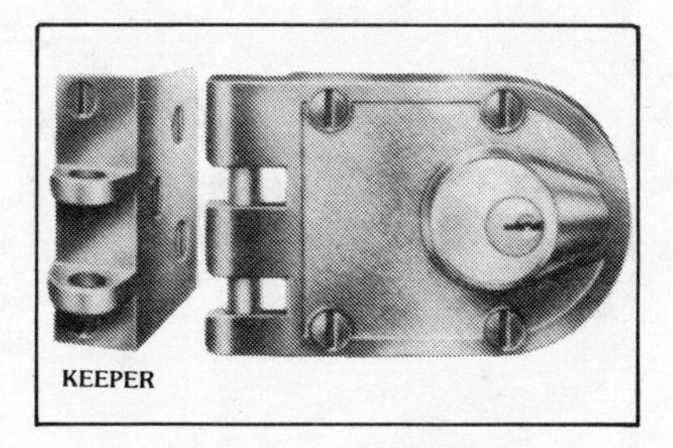

Fig. 3-5.

Not all keepers supply equal security. The one shown in Fig. 3-6A is a *flat* keeper and is held in place by only two screws. A better arrangement is the *angled* keeper, illustrated in Fig. 3-6B. This keeper is not only secured by five screws, but these screws are at right angles to each other.

While a strike and a keeper perform the same function, the keeper is more secure and is much stronger.

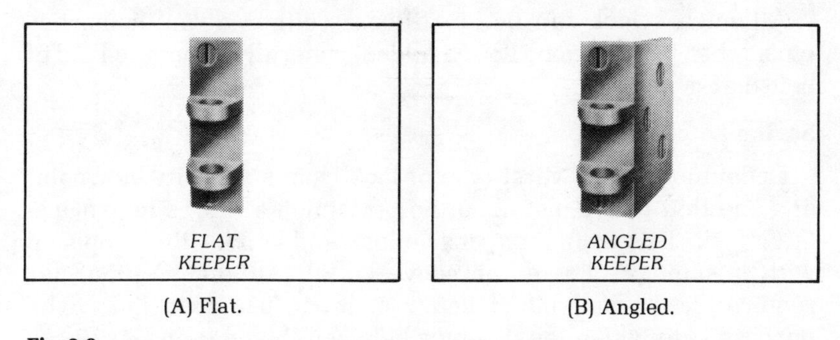

(A) Flat. (B) Angled.

Fig. 3-6.

Typical vertical bolt keepers. (Courtesy Taylor Lock Co.)

The Cylinder

The *cylinder* (Fig. 3-7), is part of the lock that receives the key, and is the mechanism for releasing the bolt so that it can be pulled back and away from the strike (for mortise locks), or the keeper (for rim locks). Both types of locks, rim and mortise, can use a cylinder. In the mortise-type lock, the cylinder is housed inside a mortise cut in the door. In the rim-type lock, the cylinder is housed in the metal body of the lock. In both mortise and rim locks, the cylinders work the same way and perform the same function.

Adjustable solid brass cylinder has screw threads for adjusting lock to fit doors 1¼ to 2½ inches thick. (Courtesy Taylor Lock Co.)

Fig. 3-7.

IDENTIFYING LOCKS

Although a lock can be identified as either a rim or mortise type, other information is often added, generally about the kind of bolt that is used.

Spring-Latch Lock

Undoubtedly, the worst type of lock from a security viewpoint and one that offers just about no protection at all, is the spring-latch lock. This is an inexpensive lock and can easily be opened with a shim (a cheater or shove knife). All that the burglar requires is a small bit of heavy metal to use as a jimmy. Its purpose is to widen the distance between the door and its frame slightly, just enough to let the shim move in. The shim can be made out of anything that is flexible, but somewhat stiff. It can be a plastic ruler, one card out of a plastic deck of cards, part of a metal venetian blind, or a small bit of scrap plastic. The burglar pushes the shim against the bolt of the spring lock, and because there is nothing to restrain its movement, the bolt moves back and the door opens.

The restraining part of the spring-latch lock is its V-shaped bolt, usually quite small, extending for approximately a ½ inch to about 1 inch beyond the lock. This doesn't mean that the entire part of this bolt is inside the keeper because there is some separation between the end of the door and the keeper. Thus, for most spring latches, the amount of bolt inside the keeper is quite small.

You can recognize a spring-latch (Fig. 3-8) lock by its bolt, which has a beveled face and can be easily moved back and forth with your finger. This doesn't mean such locks cannot or should not be used. They are economical locks for keeping the door closed against the wind.

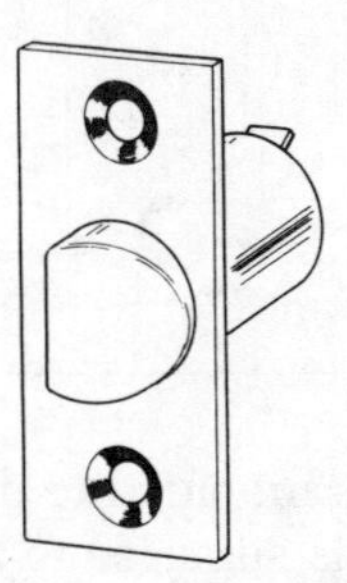

Spring latch with V-shaped bolt is adequate for inside doors connecting one room to another; but offers no security for outside entrance door. (Courtesy Schlage Lock Co.)

Fig. 3-8.

The latching-type lock can be either a rim-type or a mortise-type installation. The simplest is the rim night latch that has no cylinder and is thumb operated from the inside by turning a knob. If it is equipped with a cylinder that is mortised into the door, it can be operated by key from the outside. On the inside, however, the knob of the lock can be used to withdraw the bolt from the keeper without the use of a key (Fig. 3-9). The keeper is held in place by two wood screws.

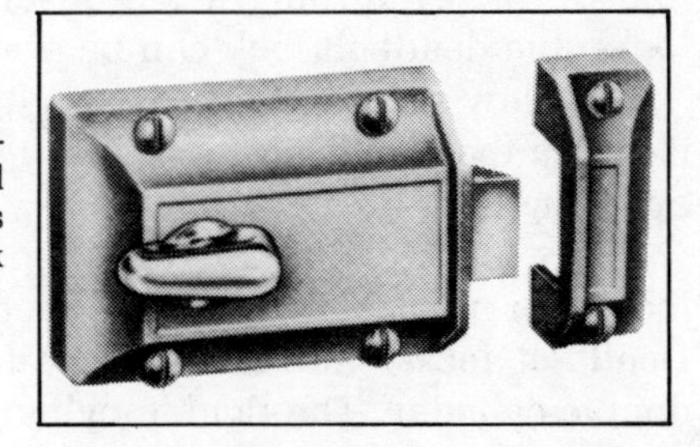

Rim night latch can be cylinder-equipped and can be locked or opened by key from the outside; inside it is thumb operated. (Courtesy Taylor Lock Co.)

Fig. 3-9.

Deadlatch (Latch Bolt)

Sometimes a short length of metal in semicircular form is combined with the beveled face bolt of the spring-latch lock (Fig. 3-10). Known as a dead latch, or latch bolt, this extra length of

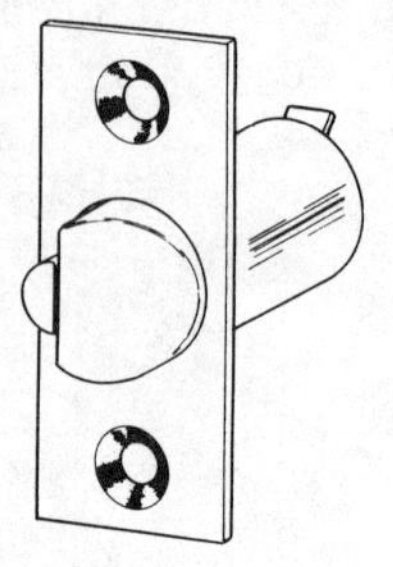

Some locks are combination deadlock and latch. (Courtesy Schlage Lock Co.)

Fig. 3-10.

metal is intended to defeat the burglar's use of a shim. This type of lock is superior to the spring-latch lock but, unfortunately, the additional bolt that is used does not have much strength, is not reinforced, and is quite short. The reason for the shortness is that it cannot extend beyond the beveled bolt and that in itself is quite short, usually only about one-half inch.

The length of the single bolt of the spring-latch lock or the combined bolts of the deadlatch can be deceiving when the lock is purchased. There is always some space between the lock and the keeper, so the effective length of the bolt is reduced. The best that can be said of the latch bolt is that it is better than the spring-latch-type lock, but not by much.

Deadbolt Lock

A deadbolt lock, as its name implies, is a lock having a bolt (or a pair of bolts) that can be moved back and forth (Fig. 3-11) or up and down, by turning a key inserted in a cylinder. Like other locks, the deadbolt lock can be a rim or a mortise type.

Most law enforcement experts agree that the best kind of lock is a deadbolt with no less than a 1-inch throw. Don't settle for anything less.

Many locks of the deadbolt type are single cylinder and are locked or unlocked by a key used on the outside of the door. Deadbolt locks, though, are available with either a single- or double-cylinder. The double cylinder is recommended for use in doors that have glass panels. An intruder may break the glass to

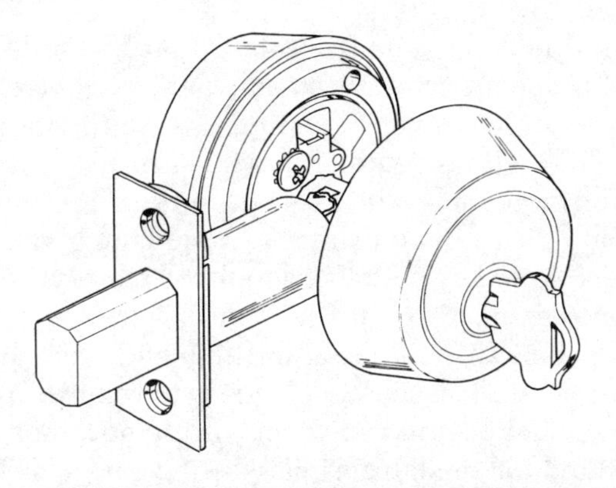

Fig. 3-11.

Mortise-type deadbolt lock.

reach the inside knob, but will be defeated by a deadbolt that requires a key to unlock the door from the inside as well as the outside.

The double-cylinder deadbolt lock can be a safety hazard if you should ever need to use the door as an emergency exit. Use this type of deadbolt only as an auxiliary lock when the house is to be unoccupied. Or else, keep a spare key close to the door, where you can use it, but where it is out of the reach of anyone trying to get in from the outside.

Another advantage of the double-cylinder lock is that it denies the use of that door to a burglar after he has ransacked your home. If it is your front door that is double-cylinder protected, you force the burglar to use some other, but less acceptable exit. He would prefer using your front door, just as you do. That exit is less likely to arouse suspicion. And, if you do keep a spare key near the door for emergency purposes, put the key in some hiding place.

The key-operated deadbolt is most often recommended, but even here you must be careful. Some deadbolts are quite skimpy. The best deadbolt is one that has a large mass—that is, is big; and has a bolt that moves a good distance into the strike. The deadbolt should extend for at least a full inch from its lock when it is opened all the way. If it measures less than this, don't bother. The purpose of a jimmy is to separate the door from the jamb. A jimmy can generally make that separation grow to about a ½ inch. A deadbolt that extends for 1 inch gives you that extra margin of security.

If you have a lock on your front door that has a deadbolt, but doesn't meet this security requirement, you have two choices. The better choice is to leave your existing lock right where it is and install a new lock with a deadbolt having a throw of not less than 1 inch. Faced by two locks, the average burglar just won't bother. Of course, you'll have the nuisance of requiring two keys (a locksmith may be able to key both locks to the same key) so that if you lose either one of them, you'll find yourself locked out. Further, you may have some difficulties about this with your landlord, so check with him first. The second choice is to remove your present lock and use the hole cut in your door for the new lock. That's the easy way out and for the burglar may be the easy way in, depending on the lock you buy.

Key-In-Knob Lock

This type lock (Fig. 3-12) consists of a lock in which the key cylinder elements are located in the knob. It is a mortise-type lock and often has a knob on the outside that is keyed and a thumb turn on the inside, but it can also have inside and outside knobs. The key-in-knob lock can use either a latch or a latch bolt. Unlike locks that offer a deadbolt only, the deadbolt used in combination with the latch not only has small dimensions, but the deadbolt section extends only for the length of the latch. For this reason, it does not offer the stronger protection of the full deadbolt type.

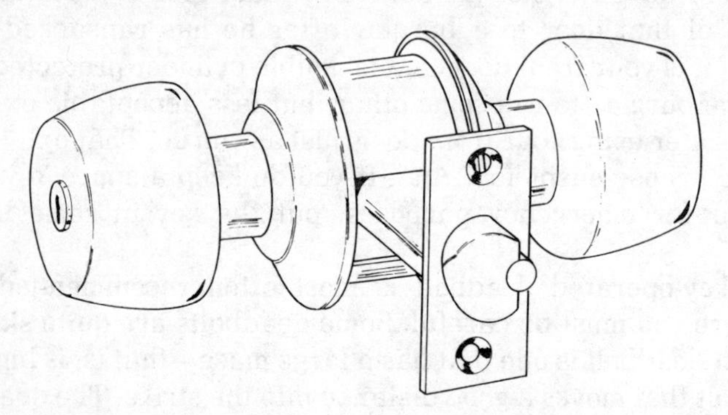

Fig. 3-12.

Key-in-knob lock with bolt that is a combined latch/deadbolt.

One advantage of the key-in-knob lock is that it is a combination unit that provides a lock plus a door handle. With other types of mortise locks, a separate handle is required, thus involving a greater expense. In the knob lock, the key-cylinder elements are located in the knob. If the knob can be cut off, wrenched off or removed by any other means, then the combined bolt/latch can be pushed back to allow the door to open.

Auxiliary Lock

An auxiliary lock is any lock that is used inside the home for locking doors. The auxiliary lock has no part that is visible from the outside and it can be opened or closed only from inside the house.

There are many kinds of auxiliary locks. Some are simple barrel-bolt types, others are deadbolt locks which may or may not be keyed. An auxiliary lock can be put on a door that is seldom used, such as a rear entry door. If the auxiliary lock is a keyed type (a deadbolt lock having a cylinder) it would be advisable to have the key for that lock placed where it is within easy reach of every member of the household. It may be necessary to use that door in case of fire. It is true that a burglar could also use that key, but here there is a difference for the thief is trying to get out of the house, not in.

If the back door on which an auxiliary lock is to be placed has glass panels, the auxiliary lock should be a keyed type. If not, it would be easy for a burglar to remove one of the glass panels, reach in, and then simply turn the lock by hand. Also, use a keyed lock if the door is a hollow construction type. With this kind of door, it is relatively easy for a burglar to saw or break through the thin front and back panels.

Thumb-Operated Deadbolt

The thumb-operated deadbolt is a good auxiliary security lock and, like all other locks, has advantages and disadvantages. The lock is a rim type and requires no key. It is thumb operated from inside the house. Thus, it is only a supplementary lock and can only be used when someone is at home. Its real function is to supply locking security for the front door when a family retires for the night.

Since it is not key operated, it does mean other members of the family will not be able to gain access to the house, possibly late at night, if the thumb lock has been turned, without ringing the front door bell. The best way to use a lock of this kind is to have an understanding among all members of the family that the last person entering the house at night will be responsible for turning the thumb operated deadbolt.

Keyless Lock

Not all locks require keys. One particular type is a pushbutton combination and instead of a key there are five pushbuttons on the outside. To unlock this device you must not only know the correct buttons to push, but you must also know the correct sequence. Any combination can be set using as many of the five buttons as you wish in any sequence you want. The buttons can be pushed individually or together with one or more of the other buttons as part of the special combination. For example, a combination of 2 and 4 pushed together, then 3, then 5, is not the same as 2, 4, 3, and 5 pushed individually and in that sequence. The lock will open only when the correct buttons are pushed in the proper order.

The lock is designed to make forced entry difficult. You can change the lock combination if you ever suspect that someone has managed to learn it. The advantage of a lock of this kind is that you can never forget your key or lose it.

THE LOCK SYSTEM

You should not buy a lock and then think no further steps are required for security. Instead, look on the lock as your initial move. You should consider the use of a reinforced strike, outside frame stripping to cover the open area where the door meets the jamb, a cylinder guard or an armor collar for protecting the cylinder of the lock, the possible use of a lock with a double cylinder, an auxiliary lock, and some kind of door system. If the door is of hollow construction it may need to be replaced with one that is solid. If the door has glass panels, consider the use of an attractive steel grille. A relatively inexpensive first step, though, is to use a cylinder guard or an armor collar.

Cylinder Guard

There are a number of ways in which a thief can defeat a lock. If the lock is a latch type, he can use a shim to separate the bolt from the strike. If the door frame is weak or if there is an available space between the door and the frame, he can use a jimmy to force the two apart, or he can use a cylinder puller to wrench or pry the cylinder out of the lock.

To protect the cylinder, you can use a brass-plated steel cylinder guard (Fig. 3-13). The plate is positioned over the cylinder and secured in place with four high strength bolts, also brass plated to match the guard. A cylinder guard is sometimes called a front plate.

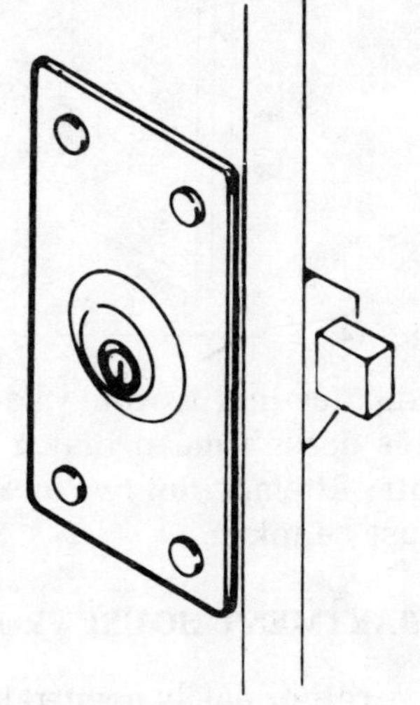

Cylinder guard. (Courtesy Taylor Lock Co.)

Fig. 3-13.

Armor Collar

Still another form of cylinder protection is the armor collar, a circular ring of tough metal that fits around the cylinder. It is free to turn and should a burglar use pincers, all that happens is that the collar turns but the cylinder remains firmly in place.

The Double Lock Advantage

For better security, it is advisable to have two locks on your front door. One of these can be a spring-latch/deadbolt type while the other can be a deadbolt type only (Fig. 3-14). To avoid the nuisance of needing to use two different keys for opening the door, both locks can have identical cylinders so the same key can open both locks.

The use of two front door locks has a psychological advantage. It warns the burglar that you are security conscious. Further, it increases the amount of time the burglar must spend in what is an exposed condition. For the opportunistic burglar, the nonprofessional who simply goes from door to door, the sight of two locks

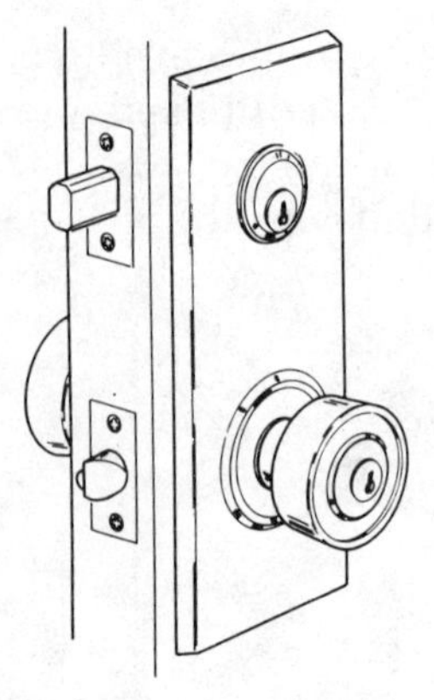

Double lock with both outside cylinders using same key; upper lock, a deadbolt type, is thumb operated on the inside; lower lock is a latch/deadbolt type.

Fig. 3-14.

indicates that it would be unlikely that both would be left open. This doesn't mean that a determined burglar may not make an entry attempt, but two locks force him to evaluate the risks which must be taken.

APARTMENT HOUSE FRONT DOOR LOCK

A rather easily opened lock is the kind that is used on the front door of an apartment house and is opened by keys supplied to all the tenants. Since so many keys are distributed, it is no great problem for a burglar to get one, but even this isn't really necessary. Such locks are also inexpensive, are manufactured to rather loose tolerances, and can be "jiggled." Jiggling is the act of inserting a key or slim bit of metal into the keyway of the lock, and jiggling the metal until it raises the tumblers. The great problem with this kind of lock, as well as with the spring latch, is that it inspires a false sense of security. With these locks, there is no security.

USING THE SERVICES OF A LOCKSMITH

There are all sorts of locksmiths, just as there are all kinds of car mechanics and television repairmen. With a locksmith, though, you literally put your apartment or house in his hands if you buy a lock from him. He has every opportunity, the equipment, and the skill to make duplicate keys to every lock he sells. They are usually honest, though, but it might be your misfortune to select one not quite as honest as he should be. When you visit the store of a locksmith, look in his window and you will find a sign indicating he is a member of some locksmiths' association. He may have a printed sign inside his shop proclaiming his membership. The sign should say also that he is a bonded locksmith. If that is the case, then you will know he has been adequately investigated. Somebody has done a checking job for you.

YOUR KEY RING

Since it is inconvenient to have single keys rattling around in your pocket, it is more practical to keep them in a small leather key case. This not only prevents individual keys from getting lost, but keeps the keys from tearing your suit lining. Some people, concerned that they might lose all of their keys at one time with the loss of their key case, very carefully put their name and address inside the case or attach a tag to it carrying the same information. Yes, they will get their keys back, probably delivered personally by a burglar. If you are concerned about losing your keys, make a duplicate set and either keep them in your bank vault or exchange keys with a trusted neighbor or relative. You hold his duplicate set and he holds yours. With favors exchanged this way, there is no obligation, and you are both carrying the burden of trust. But whatever you do, remember not to hide keys outside your home. Burglars are accustomed to looking for keys under doormats, or on top of doors. If you have an apartment, you simply do not have any outside areas for hiding keys.

CHOOSING A LOCK

Buying a lock can be troublesome since there are so many to choose from at such a variety of prices offered. But if you are looking for security, and you should be, eliminate all spring-latch locks or spring-latch locks equipped with a deadbolt. Get one with a deadbolt that is as thick and long as possible. Not all deadbolts are alike. One that has a core of hardened steel is preferable.

Examine the face of the lock. The lock should be tapered from the face to the rear—the part that will rest against the outside frame of the door. If this part of the lock is tapered, a twisting tool such as a wrench will not be able to rotate the lock off of the door. This part of the lock is called the cylinder guard and it should be made of solid, heavy metal.

There should be no screws on any part of the lock that faces the outside. Obviously, such screws could be turned by anyone having a screwdriver. If your lock does have one or more screws that are on the outside of the lock, whether single-slotted or Phillips-head types, damage the slots so that no screwdriver can be used to turn out the screws.

LOCK PICKING

Manufacturers of locks may sometimes refer to their locks as **pick resistant,** but this is a vague generality. Every lock is pick resistant. The difference is in the amount of resistance. There is no standard against which so-called pick resistant locks can be evaluated or measured.

Picking a lock requires a certain amount of skill. It is doubtful if a burglar equipped with the expertise to pick a lock would waste time and effort working on picking a household lock. If a burglar has enough on-the-job experience, he or she can usually look at a lock and, knowing its weakness, force an entry.

DOORS

Burglars prefer front door entry, since this is the customary way of getting in and out of a home. For the burglar it is less likely to arouse suspicion, particularly if one can get in and out quickly.

Door Fit

The door should make a good fit and not have a gap between it and the door jamb. Even a small space will permit the use of a crowbar. A good strong lock is no better than the door on which it is mounted. The maximum space between a door and the frame should be 1/8 inch, preferably less. If the gap is too large, you can add trim, preferably metal, to the outside of the door to cover the gap.

Many entry doors made today are of hollow construction. You can easily check to see if you have a door of this type by rapping on it. If this produces a thud that does not change as you tap various areas of the door, then it is solid. A hollow door will produce a characteristic hollow sound, except around areas near the edges where the sound will be a thud. The best and safest arrangement is to replace a hollow door with one that is solid with core wood at least 1-¾ inches thick. Metal entry doors are also available. These are preferable since they do not cut through as easily as wooden doors, however, they are also more expensive.

Glass Panels

If the door is a type that has glass panels, but is solid otherwise, reinforce the panels with a decorative metal grille. A glass panel, especially if it is positioned near the lock, is the weakest area in the lock/door combination. The glass can be cut or smashed and the burglar can get into your house just as fast as you can with a key. You can also use a high strength plastic sheet, putting this in place directly behind the glass. Neither the metal grille nor the plastic sheet will reduce light input to any great extent.

The metal grille or the plastic sheet should be kept in place with screws. To keep the screws from being removed, fill in the screw slots with wood dough. Another method is to file the screw heads so that the screw-head slots are no longer usable. Some hardware stores also carry burglar-proof screws. These permit the screws to be turned in, that is, fastened, but not to be turned in the other direction.

Door Jamb

If the door jamb in the area of the lock has rotted or is weakened, have the jamb replaced. You can easily determine the condition of the jamb by unscrewing the strike. If the screws holding the strike in place turn too easily, or if you can keep turning them clockwise indefinitely with a screwdriver, then the jamb is no longer supporting the strike properly. In some instances, where the jamb has weakened and will no longer hold the strike, some homeowners remove the strike altogether. This is practically an invitation to theft since it now forms the weakest part of the lock system. You can replace the strike with one that is a reinforced type, a device that will take some of the pressure off of the jamb.

Door Opens Outward

Front and rear doors generally open inward. The advantage of this arrangement is that the side of the door covers the door hinges when the door is closed. There are some homes in which the door opens outward. This means that the hinge and hinge pins are exposed. Such a door can easily be opened by removing the pins. Have your local locksmith replace them with nonremovable hinge pins.

Door Chains

Door chains as shown in Fig. 3-15 were mentioned briefly in

The best front door chain lock is one that is key operated.

Fig. 3-15.

Chapter 2. There are two types—unlocked and locked. The unlocked type is cheaper, however, the locked type provides more security. A burglar can use a bent coat hanger to move the usual type of unlocked door chain out of the way. With a lock-type door chain, after a burglar has managed to defeat the regular door lock, the door opens just a few inches and cannot be opened further unless the lock on the door chain can be defeated. A cheap door chain can be easily snipped; the better grades use hardened metal requiring a powerful tool for cutting. The lock-type door chain is excellent for use with doors that have glass panels. A burglar that removes a pane of glass only to be thwarted by the chain lock may have second thoughts about jimmying the lock.

When you buy a lock, consider it the way you do insurance. You aren't buying a lock; you are buying protection. A quality lock, properly installed, should last the life of the house. If you select a bored lock—either tubular or cylindrical—for a new door or to replace a worn-out lock, you should be able to install it yourself if you have average mechanical ability and the tools normally found in a home shop. However, if you have special locking problems—for example, your front door may be metal instead of wood, or it may have an artistic design—it would be best to locate a firm that specializes in builder's hardware to advise you. Or, you may want to consult a locksmith who qualifies as a security expert or master locksmith. Security consultation and inspection will probably be included in the services as well as selling and installing locks.

You can also get a door chain equipped with an alarm. The same chain/alarm unit can also be mounted on a window. Opening the door or window pulls a chain which, in turn, sets off the alarm that is operated by a 9-volt battery.

Knob-Type Chain Guard

The knob-type chain guard shown in Fig. 3-16 is somewhat different than the usual door guard. The guard is held in place by a 3-inch anchor bolt. The ring, at the opposite end of the chain, slides over the door knob. The heat-treated case-hardened chain cannot be sawed. The lock provides added security by preventing removal of the ring from the outside of the door, but it is easily removed from the inside. This type of guard does not use a key and is effective only when someone is inside the house.

Fig. 3-16.

Knob-type chain guard. (Courtesy Taylor Lock Co.)

PATIO DOOR

Because they have a large expanse of glass, patio doors are often a weak area in the security arrangement of a house. The locks that are supplied with patio doors are weak, and often aren't even functional. It is no great difficulty for a burglar to cut away enough of the glass to be able to reach the lock.

Some patio doors are single-glass types and these are the easiest of all to break into. A better type is the patio door that uses two panes of glass with an air space between them. This offers a little better security and is also better from the viewpoint of losing less heat in the winter. If you aren't sure whether your patio door is single or double glass, light a match in front of the glass. If the glass is single, you will see just one reflection, if it is double, you will see two images of the match.

Another problem with patio doors is that they usually have a panel using screen mesh. One glass section is pushed to one side and replaced by the screen mesh panel to supply ventilation during the summer. The mesh is even easier for a burglar to cut through than the glass.

A patio door is really a two-part door consisting of one glass panel that is fixed in position and the other panel being allowed to slide back and forth. The sliding panel should be inside the house, the fixed panel outside. Oddly, sometimes this arrangement is reversed. If the movable panel is on the outside, then the installation is incorrect. With this setup, it is no great difficulty for a burglar to force the panel to move.

There are a number of ways of keeping a burglar from getting in via the moving panel. One method is to use a Charley Bar, a hinged metal bar that effectively keeps the moving panel fixed in position. The problem with the Charley Bar is that it is an all or nothing at all device. When the bar is pushed up and out of the way the sliding patio door can be opened to any position, and so the bar offers no protection when ventilation is wanted. However, it is superior to the lock supplied with the patio doors.

As an alternative to the Charley Bar, you can put a cut-down broom handle in the door channel. A suitable length of steel rod or shower curtain rod will also do. These should be cut to be shorter than the width of a door panel so that they can be easily lifted out of the door channel when necessary.

Wedge for Patio Door

One way of partially opening a sliding patio door is to use a rubber wedge in the track. The problem is to find a wedge of the right dimensions to fit. You can get a wedge-shaped block equipped with a Velcro® pad instead of the rubber wedge. When inserted in the track, the item shown in Fig. 3-17 fits on the frame and is held in place by a Velcro® fastener. It can be removed easily when required and does not need any drilled holes.

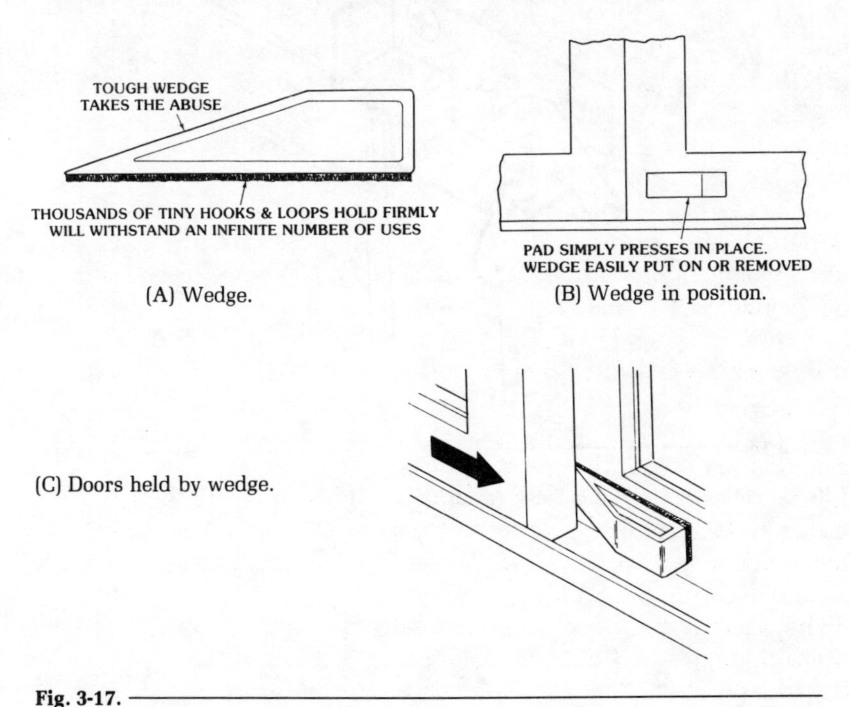

(A) Wedge.

(B) Wedge in position.

(C) Doors held by wedge.

Fig. 3-17.

Wedge-shaped block with Velcro® pad permits limited opening of patio door for ventilation. (Courtesy Taylor Lock Co.)

Keyed Patio Lock

The keyed patio lock shown in Fig. 3-18 is designed for installation at the top of the patio-door frame. It features keyed locking bar operation and locks both patio doors directly to the frame. A hardened-steel backplate mounts behind the frame for maximum security. This lock can keep the doors in a closed or partially open ventilating position.

Patio-Door Pin Lock

You can supplement the lock supplied with your patio door with a simple, inexpensive unit known as a patio-door pin lock. As shown in Fig. 3-19, it is simply a pin that fits through the top end of

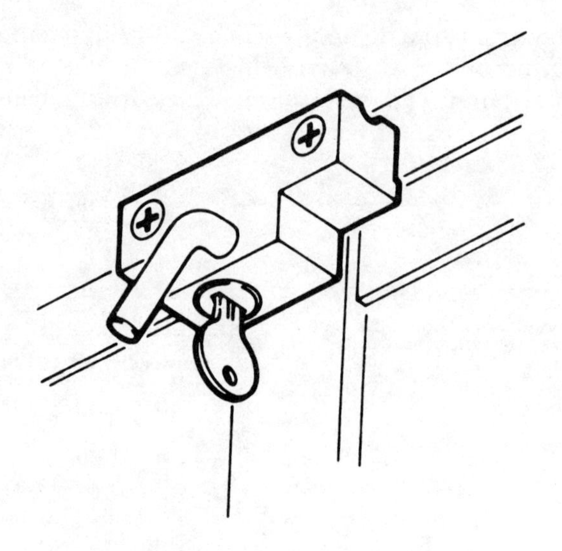

Fig. 3-18.

Keyed patio door lock. (Courtesy Taylor Lock Co.)

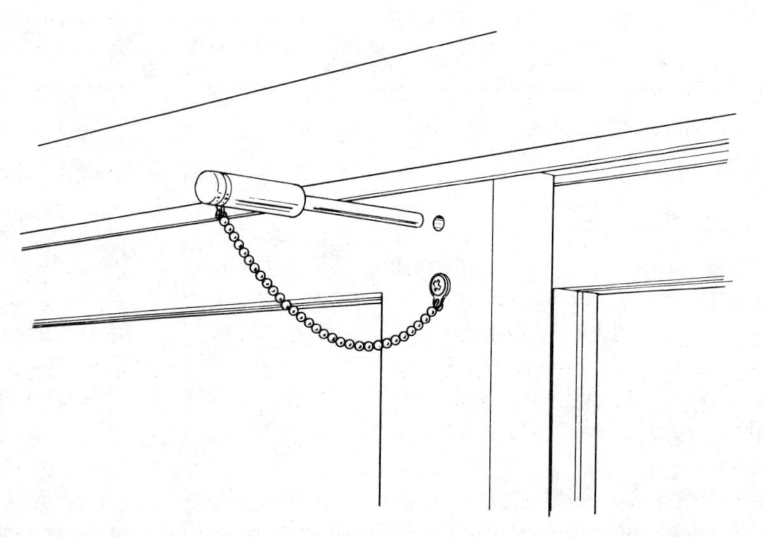

Fig. 3-19.

Patio door pin lock. (Courtesy Red Thumb Home Centers, Causeway Lumber Co.)

the fixed and sliding doors. To use the pin, it will be necessary to drill holes in both doors.

The disadvantage of this lock is the same as the original patio-door lock. The pin can only be used in one position and so does not supply security when the patio door is opened for ventilation. If a burglar cuts the glass, he can easily withdraw the pin.

Security for patio doors is just as important as for the front door, and possibly even more so. Patio doors are often at the rear in an area receiving little light and sometimes obscured from the view of neighbors. An open patio door also supplies a large exit area, something the burglar needs when removing large objects, such as a television set or microwave oven.

Garage Doors

Any door, whether front, rear, or garage, is an area of possible entry for a burglar and so requires some security attention. While the front door is the prime choice for the burglar, there is no objection if the garage door is accessible, and it frequently is.

If the garage can be reached from the inside of the house via a door, that door should have an auxiliary lock that can be opened only from the inside of the house. Thus, if burglars gain entry into the garage, they must still overcome the auxiliary lock. Unfortunately, this still gives the burglar protection of the garage—that is, they need not work out in the open where they can be observed. At the same time, however, the auxiliary lock does require time to overcome. The inside door leading to the garage should not be a hollow-panel type, but should be solid. The door should also be covered with metal sheeting on the side facing the garage to supply a fire barrier.

A garage door with glass panels is more subject to possible break-in than a solid door. This means such doors require extra protective mesures. You can cover each of the glass panels with a sheet of thick plastic that can be screwed into position from the inside. This will cut down somewhat on the amount of light coming into the garage, but it does offer greater security.

There are two basic kinds of garage doors. The first, quite common at one time, is mounted on hinges along the side of the door. The door is in two sections, with each section capable of being opened independently. Quite often, the doors are kept closed by a simple, lever-type lock at the center. This hardware is not really a lock but is just used to keep the doors closed.

The type of lock to be used for such a garage door depends on whether there is access to the garage from inside the house or not. If there is no access, then the garage door must be locked from the outside. A simple method is to use a good quality, hardened-steel hasp and a strong padlock.

The same technique can be used if the garage door can be reached from inside the house, but the advantage is that there are more security options. Thus, you can use a cane bolt, as shown in Fig. 3-20, or a surface bolt (Fig. 3-21) to secure the door. The

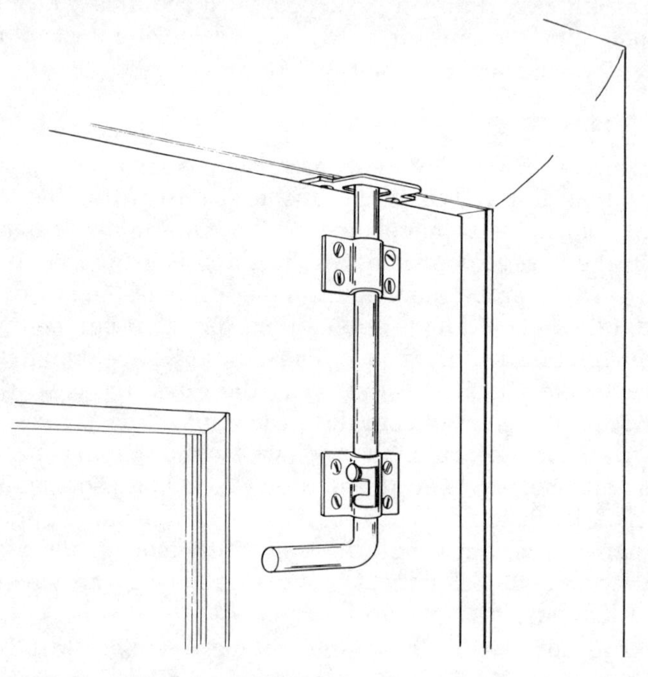

Fig. 3-20.

Cane bolt can be used to secure garage door; does not require key.

advantage of the cane bolt and the surface bolt is that no key is needed and either bolt can be unlocked just by pulling down on it. If the garage is a two-door type, that is, has doors for two cars, use a pair of cane or surface bolts, one mounted on each door.

The disadvantage of either the cane bolt or the surface bolt is that if you can open them easily, and you can, then they present no problem to any burglar that can get at them.

These two types of bolts can only be used if there is access to

Mount surface bolt strike as close to bolt as possible. (Courtesy National Manufacturing Co.)

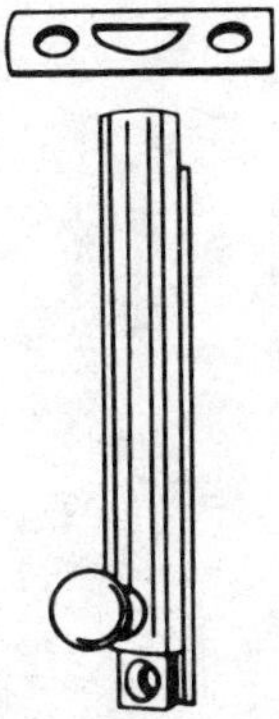

Fig. 3-21.

the garage from the house. They are helpful in that they deny access to your home via the garage. But once the burglar has entered your home and is able to get into the garage, they have no effectiveness. A more positive security arrangement would be to use a locking method that would not only deny the burglar the use of the garage door (or doors) as a means of entrance, but would also keep them from being used as an exit. Further, if your car is parked in your garage, securely locked doors will keep the burglar from considering your car as a means of transporting your possessions.

On garage doors, you can take advantage of any of the locking methods used on front or rear doors of a house or apartment; either thumb-operated or key-operated deadbolt locks will work well. A latch-type lock, however, is unless.

If the garage uses a manually-operated rolling overhead door, an easy way of getting security is to use a length of chain, putting the chain through the track on which the door rides. As an added measure of security, connect the open ends of the chain with a sturdy padlock. To avoid the nuisance of carrying the extra key with you, hide the key somewhere in the garage where you can find it easily, but hidden from a burglar. If you have a workbench, for example, mount the key on a nail on the underside of the table or use a cup hook. Just make sure the key isn't visible. If you don't wish to be bothered with a chain, you can use a padlock with a long shackle to secure one door, as shown in Fig. 3-22. Fig. 3-23 shows how the padlock is mounted on the inside roller track of the door. Two doors can be held with only one padlock by using a length of chain through both tracks.

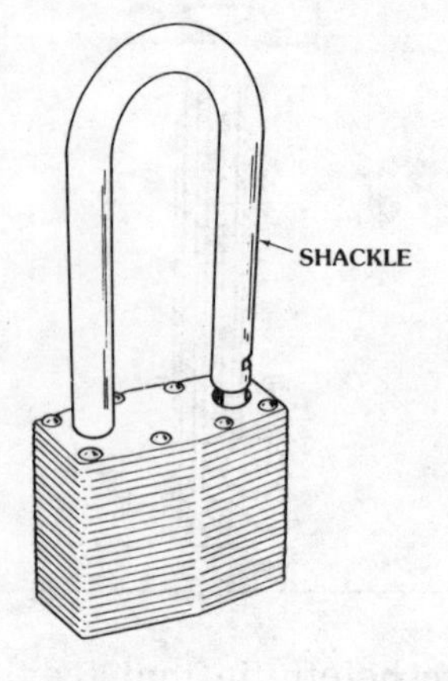

Padlock with long shackle. (Courtesy National Manufacturing Co.)

Fig. 3-22.

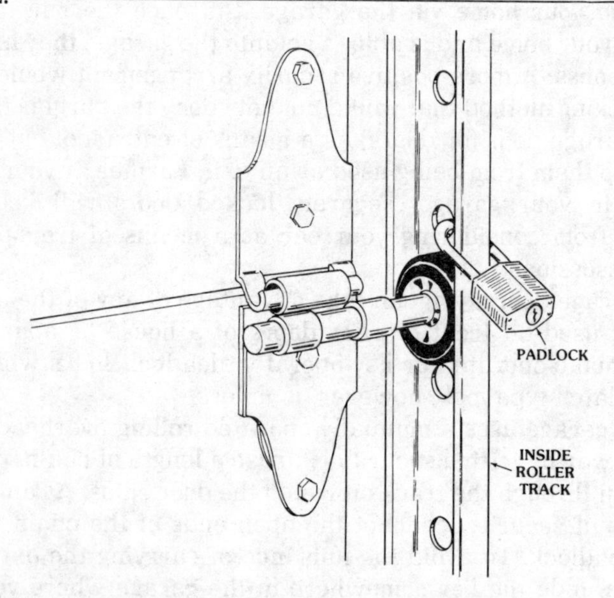

Fig. 3-23.

Padlock can secure overhead rolling garage door by mounting it on inside roller track.

Radio-Controlled Door

If you have a rolling overhead garage door (or doors) you can use a radio-controlled garage-door opener. There are two basic types—single frequency and dual frequency. The disadvantage of the single frequency type is that a signal produced by someone else's garage door opener may accidentally open your garage door. A spurious signal from any other source may also do the same thing.

A better arrangement, if you want the convenience of an electronically controlled garage door, is to use a dual-frequency device. With this electronic door opener, two signals are needed and the signals must be in the correct sequence. This arrangement not only offers security but convenience as well, since the garage-door opener is operated from your car. The disadvantage is that the dual-signal electronically controlled garage-door opener is more expensive than the single-signal type.

PADLOCKS

The kind of padlock that you should buy depends entirely on the purpose for which you want to use it. If you need such a lock for a drawer in a desk in your home with the idea of keeping prying fingers out, then an inexpensive padlock will do. But if you are going to use a padlock to protect a bicycle, a moped, or a motorcycle against theft, then you need one that will supply protection.

For security, a padlock should be made of case-hardened steel and that word, "hardened," should appear somewhere on the lock. It will probably be stamped right into the metal. Case hardening gives an extra-hard outer layer that resists cutting or sawing, and a tough inner core that keeps the shackle from becoming brittle. The shackle (see Fig. 3-24) should have a minimum dimension of 9/32 inch.

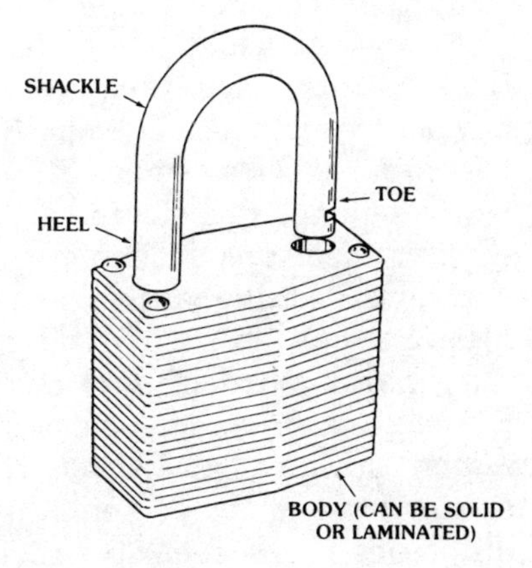

Main parts of a padlock. (Courtesy National Manufacturing Co.)

Fig. 3-24.

The padlock should have a double locking mechanism, with a lock for the heel as well as the toe and it should also have a five-pin tumbler. In better grade padlocks, there will be a key holding feature, with the lock retaining the key until the padlock is in its fully closed position.

On some padlocks, you will find a key number somewhere along the base of the lock. The fact that such a number is supplied does not of itself make the lock a good one or a poor one, but as a general rule, such a number is only put on better grade locks. The purpose of the number is to let you get a duplicate key made in case you lose the original. The disadvantage, of course, is that anyone can copy the key number and have a duplicate key made by a locksmith. For such locks, copy the key number and file that information together with your important papers. Then remove the key number from the lock, even if this involves filing the number off the base.

Wrought Steel (Shell) Padlock

Wrought steel or shell padlocks are the lowest priced locks available. They are a good buy if you understand that they are intended mainly for nuisance protection—keeping children out of your tool box, locking power tools against tampering, restricting access to a mailbox, storeroom, etc. Costing little, they can prevent thoughtless misadventures or injury from hazardous household and industrial items.

Warded Padlocks

Warded padlocks are of laminated construction that makes them sturdier than shell locks and gives added resistance to anyone trying to smash them. A warded locking mechanism does not permit as wide a range of key combinations as those available with pin-tumbler mechanisms. A major use for this lock would be where property is of limited value, such as oil sump tank caps, well covers, duffelbags, farm gates, beach lockers, etc.

Because of the relatively large clearances between its operating parts, warded locks are frequently used in applications where sand, water, ice, and other contaminants are encountered. A word of warning, however. While they look almost the same as master pin-tumbler padlocks, at first glance, there is a decided difference in the security provided. The most visible difference with a warded lock is in the type of key employed. Because warded padlocks cost about half as much money and because, outwardly, pin-tumbler and warded locks look so much alike, it is easy to make the wrong padlock buying decision where high security is essential. When the property to be locked has substantial value, the pin-tumbler lock with its many hidden strengths is infinitely superior and is well worth the additional cost.

Pin-Tumbler Laminated Padlocks

Also known as master locks, and controlled by a precision mechanism, laminated pin-tumbler padlocks offer premium protection and also the possibility of thousands of key combinations. A warded-lock mechanism has only four working parts while a master pin-tumbler lock has nineteen. One such lock made by Master Lock Co. has a patented double-locking feature that multiplies this protection by independently locking each shackle leg (heel and toe). The result is a lock that is extremely difficult to open by forcing, shimming, or rapping.

Corrosion Proof Padlocks

Hard-wrought-brass versions of master high-security laminated padlocks are available. They are designed to withstand severe corrosion problems such as those encountered along seacoasts, aboard boats, at oil refineries, and in areas of high humidity and atmospheric pollution. For burglary deterrence, the shackle typically will be chrome-plated hardened steel. However, the price for maximum security brass padlocks is substantially higher than for steel.

A lower cost medium-security alternative for the average user is the solid-brass padlock. Again, look for pin-tumbler locking and a case-hardened shackle to protect valued property. Priced substantially less than heavy-duty laminated brass padlocks, solid-brass locks are naturals for many uses where corrosion is a possible problem—boats, outdoor lockers, gates, and similar applications.

Combination Padlocks

The primary reason for using a combination padlock (Fig. 3-25) is its convenience—it doesn't require a key. It is particularly helpful with children where lost keys may be a problem. Protection features to look for include reinforced double-wall construction, a tough stainless-steel outer case over a sturdy wrought-

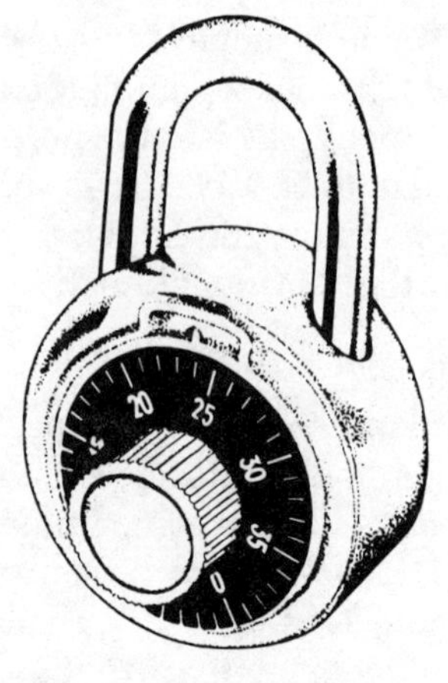

Combination padlock. (Courtesy National Manufacturing Co.)

Fig. 3-25.

steel inner case, and a hardened-steel shackle for added resistance against cutting and sawing.

The Hasp

It is essential to consider more than just the strength of the padlock. The finest lock affords little protection if burglars find it hung from an undersized or unhardened hasp which they can cut with ease. To avoid a weak link in security, get a hasp (Fig. 3-26) that matches the quality of the lock. Look for adequate size, a pinless hinge, concealed screws, case-hardened staple (the metal loop the lock passes through) and steel ribbing for added strength.

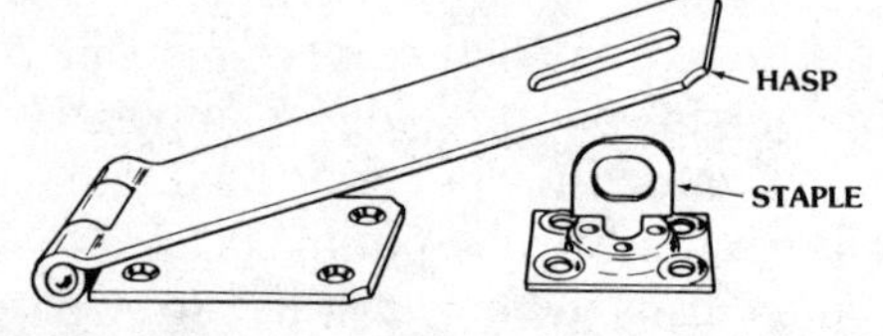

Screws holding hasp and staple in place are covered when hasp is closed.

Fig. 3-26.

Hasplock

An evolutionary step beyond the hasp is the hasplock (Fig. 3-27). Here, instead of using a separate padlock, the lock and hasp are one. Permanently joined so that the lock can't be misplaced or stolen, hasplocks give built-in convenience similar to the deadbolt door lock, while offering much easier installation. Their uses range from garage doors and sheds to boat houses, trucks, and warehouses.

Improved type hasp and staple works as positive door latch as well as locking device. When hasp is pushed over staple, it is automatically latched to prevent movement of door—will open with only one hand. (Courtesy Stanley Hardware, Div. The Stanley Works)

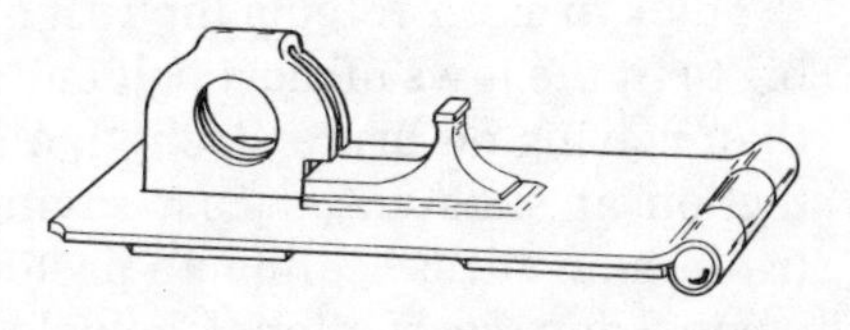

Fig. 3-27.

Choice of Chain or Cable

When it comes to movable property, the use of a strong padlock with a properly matched chain or steel cable is the proper method. This combination enables owners of bicycles, mopeds, motorcycles, boats, and other easily stolen goods to obtain maximum protection by securing them to a post, tree, dock, or other immovable object.

Common chain is regularly available in hardware stores. But don't bother. Look for a chain specifically designed for locking applications. Be sure it is case-hardened for high resistance to bolt cutters, saws, and files. The individual links should be welded, not just twisted, to resist being pried apart.

Multistranded security cable is available for equivalent protection, with the added benefit of light weight. In either case, the thicker the chain or cable, the greater the protection. Examine the cable closely because some manufacturers add a thick coating of vinyl to make a small steel strand look bigger.

For greatest protection, position the lock and cable (or chain) as high above the ground as possible. This makes it difficult for the thieves to gain extra leverage by bracing one leg of a bolt cutter against the ground.

SPECIAL PADLOCKS

You can also get locks that are specifically designed for a particular purpose.

Armorlock

In the contest of thief versus property holder, the Armorlock advanced padlock introduces a special shackle to frustrate thieves armed with hacksaws and bolt cutters. This padlock is designed with these cutting tools in mind. It has a free-spinning shackle sleeve of case-hardened steel that turns like a roller bearing so that a saw blade will slide over its surface instead of biting in.

Thick armor protects the rest of the shackle and it is simply too big to fit the jaws of most bolt cutters. By outwitting thieves rather than relying on brute strength alone, Armorlock gives extra protection at substantial cost savings. Among its uses are locking trailers on cars; beefing up protection for bicycles, snowmobiles, cabins and recreational vehicles; securely anchoring valuable portable equipment; and locking up other prime theft targets.

Gun Lock

This lock is a life saver as well as a theft thwarter, and it is particularly important in homes where there are small children. The gun lock blocks access to the trigger, stopping anyone from firing rifles, shotguns, or handguns inadvertently. Unobtrusive, they can even be used for firearms on display. A lock on a gun also acts as a deterrent for the burglar who is ransacking a home because the gun with a lock on it is of no value to him.

Trailer Lock

This specially designed lock for trailers guards against towaway theft of a trailer and its contents. Too many people leave trailers with expensive boats, snowmobiles, etc., unhitched and unguarded in their driveways—easy targets for thieves. A master trailer lock completely blocks access to the coupler cavity and it cannot be pried off.

Outboard Motor Lock

This lock secures the motor to the transom of the boat by making it impossible to unscrew the clamp handles. Too often, theft of an outboard takes little more effort than stealing an unlocked bicycle. These motors are among the most poorly protected pieces of property people own. A master outboard-motor lock provides the security of case-hardened steel combined with pin tumbler locking. It defies cutting by saws or bolt cutters.

Ski Lock

This lock is a way of fighting back against the theft of skis and poles, a serious and growing threat to sports enthusiasts. The ski lock has a master combination-lock mechanism with an oversize dial that can be operated easily even when wearing gloves. You can use a suitable cable length to secure both the skis and poles to a tree, rack, or any other suitable immovable object. The lock is designed so that the owner can carry it effortlessly around the waist while skiing.

Emergency Breakaway Padlock

This lock, designed to prevent nuisance tampering with fire equipment, speeds fire emergency action. It can be opened instantly and without a key in an emergency. At the blow of a hammer or wrench, its frangible shackle breaks along previously built-in indentations. This lock offers fire safety specified by top insurance underwriters for sprinkler valves, fire hose housings, and other applications where fast, easy access is vital.

IF YOU ARE IN DOUBT ABOUT A LOCK

The best place to seek answers to questions of security is your local police department. Most law enforcement agencies have experts to help you and may even have displays of security locks, including both mortise and rim types, plus an assortment of padlocks.

When in doubt and unable to get expert advice, the rules to know are: (1) Buy the best padlock protection you can afford (strength and cost generally coincide). (2) When possible, avoid leaving locked items in out-of-the-way places where thieves have time to work on the lock unseen. (3) If locked property is of a nature that it has a strong appeal to the light fingered, back up your locks with some insurance. (4) Consider the use of supplementary deadbolt locks on all outside access doors to your home. (5) Talk to your locksmith about reinforcing the strike on your door. If the area around the strike on your door is rotted, consult a carpenter about replacing the entire jamb, or, if you are a competent do-it-yourselfer, be aware that a weakened strike means a weak locking system. (6) Consider the use of double-cylinder locks for doors equipped with glass panels. (7) Quite a few people rely on locks for protection, and then buy the cheapest, poorest-made lock. A good lock is one that has a deadbolt of ample size, whose cylinder cannot be pulled out, and is sturdy enough not to be punched out. A good lock is one that requires a key (or some other inserting device) to open it and to close it, and that also has an additional deadbolt operative only from the inside. A good lock is one for which only the home owner or apartment dweller has the key, and it is *not* the lock inherited from the previous occupant. A good lock is one in which the strike is completely covered on the outside. A good lock is mounted on a door so that the edge of the door and its jamb are covered on the outside. A good lock is also mounted on a door whose hinges cannot be removed from the outside. A good lock is one that is supplemented by a *strong* key-type chain lock. Unfortunately, many people buy locks on the basis of price. Finally, a good lock is one that is installed by a professional or master locksmith who has been advised that security is of paramount importance.

Chapter 4

Alarm Systems

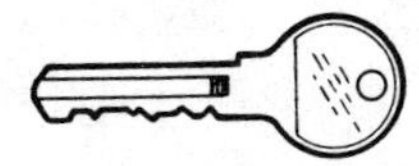

According to FBI statistics, only one in forty-one thousand intruders will remain when an alarm is sounded (Fig. 4-1). They

An unexpected alarm, indoors or out, is often enough to deter the potential burglar.

Fig. 4-1.

all fear detection. There are any number of alarm intrusion devices you can use, from those that set off an alarm if a window is open or broken to those that sound with the opening of one, or more, doors in your home. Some use electronics, depending on a transmitted signal, while others are completely wired systems. One type uses an ordinary-looking mat that can be placed near the front door so that any intruder steps on it automatically. Or, it can be hidden under a rug and is activated when the burglar walks across it. The mat is a pressure-sensitive device, containing a switch wired to a battery-operated alarm.

The best deterrent for burglars is to have an occupied home, but even such a home is more secure with pick-resistant and jimmy-resistant locks installed by a professional locksmith. But if a home is not to be a prison, it must be a place you can leave with reasonable peace of mind and so it makes sense to have a sensitive intrusion alarm system. More popularly called burglar alarms, such systems offer continuous protection.

ELEMENTS OF AN ALARM SYSTEM

Regardless of their construction, design, application, or sophistication, all alarm systems have three functional elements: the **sensor, control,** and **alarm** (Fig. 4-2). They may be separate devices or may be contained in a single housing, but all three elements must be present for effective operation.

Fig. 4-2.

Basic alarm system.

The Sensor

The sensor, as its name implies, "senses" or detects a condition. Depending on its design, it may detect fire, smoke, a drop in temperature, humidity, sound, light, pressure, or movement. It can be as simple a device as an electrical switch or as complex as a tv camera.

The sensor works because of some action by the burglar. The burglar, by the pressure of walking, may close a switch, may break through a light beam, or may disturb an electrical-field pattern set up in your home. Whatever the sensor may be, the burglar causes it to sense, or become aware of, some change in conditions. And when this happens, the sensor sends a signal to a control device. The control then sets off the alarm. In the simplest of systems, the sensor and control can be combined in a single unit.

The Alarm

The alarm can be any one of a variety of devices (Fig. 4-3). It can be a noisemaker such as a bell, buzzer, siren, or horn. It can

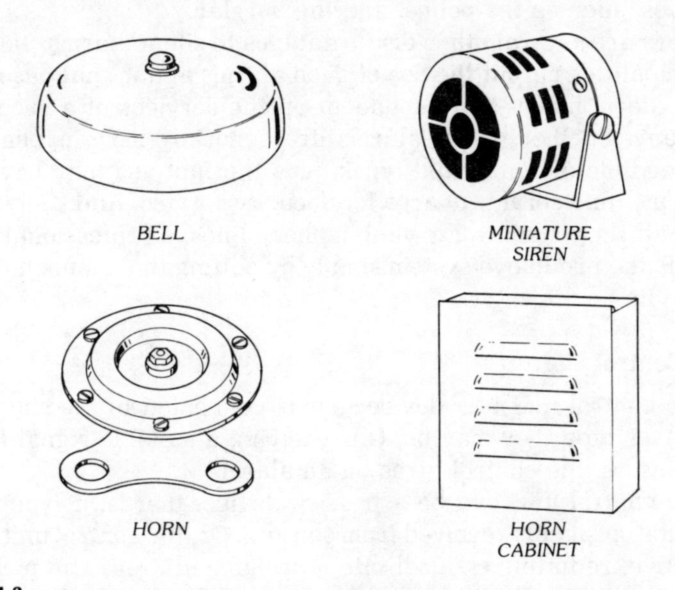

Fig. 4-3.

Typical alarm devices. (Courtesy Hydrometals, Inc.)

be a warning pilot lamp or a floodlight. Some installations make use of what is known as a silent alarm. Instead of using a noisemaker, the system is connected through a leased telephone line or by a broadcast radio signal to either a police substation or the central office (sometimes called the central station); or to a burglar-alarm company or detective agency that either dispatches guards or calls the police. Unfortunately, central-station alarm systems are not suitable for do-it-yourself installation.

Performance approaching that of a central-station alarm system can be achieved with a self-installed intrusion alarm, if the home owner uses a commercially available instrument known as an automatic telephone dialer. In operation, the instrument is connected to the alarm system and, in case of intrusion, will dial a predetermined number and play a pre-recorded message.

Aside from the way they function, there is a difference between the silent alarm and the kind that produces a warning sound or lights. The basic function of the silent alarm is to apprehend the criminal. This is no problem if you are away from home at the time the silent alarm is tripped. But if you are at home, then you put yourself on the scene of a confrontation between security services, such as the police, and the burglar.

There are several other disadvantages to silent alarms. Not all municipalities permit the use of such alarms and in that case the silent alarm makes you dependent on the services of a security company. Further, silent alarms are generally more expensive than audible and/or visual types, plus the fact you may have to pay a monthly service charge for security service. And, since the silent alarm system works via telephone lines, a professional can disable the protective system simply by cutting the telephone line before he forces entry.

The Control Unit

The control unit is a device that is connected to the sensors, whatever type they may be. On receiving a suitable signal from the sensors, the control turns on an alarm.

The control unit can be a passive device, that is, a type that depends on signals received from sensors. Or, the control unit can be active, radiating a signal into a designated area, and picking up the same signal. Thus, this type of control unit is both a transmitter and a receiver, and so is not dependent on external sensors. It works as a combined sensor and control. In either case, it is still the control unit that turns on the alarm, usually by permitting the flow of an electrical current to a bell, a horn, or lights.

FORMS OF ALARM PROTECTION

Depending on its design and installation, an intrusion alarm system may provide for any of several different forms of protection as shown in Fig. 4-4. It may serve, for example, to detect an intruder trying to enter a protected area. Or, the system may respond to an unauthorized person anywhere within the premises or in a specific room. Finally, the system can be designed to respond only if the intruder approaches or touches a specific object, such as a single door, a safe, or a file cabinet. Sometimes a single installation can be set up to furnish two or more forms of protection simultaneously.

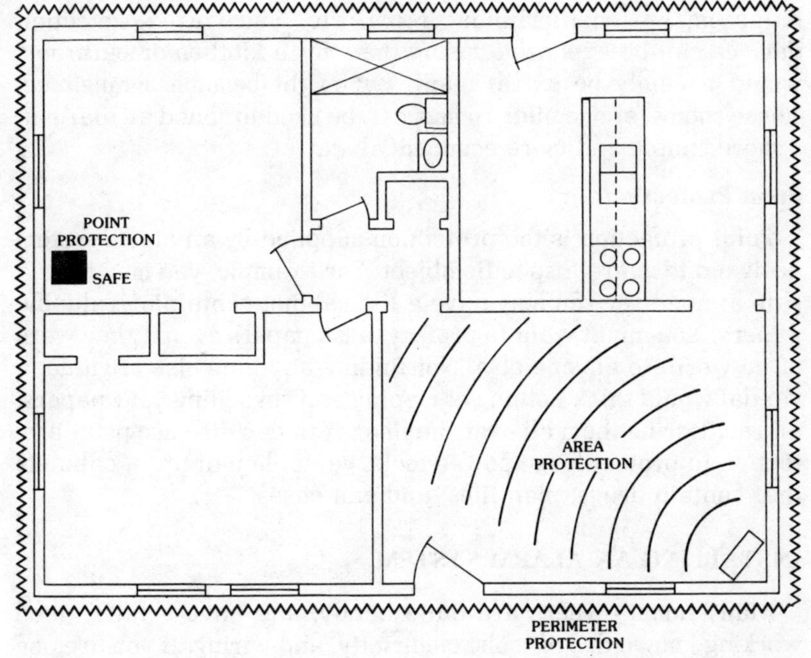

Fig. 4-4.

Three basic forms of protection—perimeter, area, and point. (Courtesy Hydrometals, Inc.)

Perimeter Protection

Consider a typical small house. If an alarm system protects all the doors, windows, and other entrances to the home, or the entire exterior, it is said to offer **perimeter** protection. In a physical sense, a fence around a piece of property is perimeter protection. But you can also have perimeter protection in other ways. Floodlights used on the grounds around a house supply perimeter protection. Perimeter protection systems can be used to guard a specific area within a building as well as an entire building, as, for example, the entrances to a stock room at a restaurant or bar.

Area Protection

If an alarm system can detect an intruder anywhere in a given location, such as the living room in your home, it provides **area** protection. You can set up several such systems to protect those rooms containing your valuables, ignoring other rooms, such as bathrooms, kitchen, laundry room, etc. Since area-protection devices can be expensive, using them for a kitchen or bathroom would not only be extravagant, but might become a nuisance. These rooms, and similar rooms, can be used to sound an alarm in a more simple and more economical way.

Point Protection

Point protection is the protection supplied by an alarm system designed to guard a specific object. For example, you may have a safe at home, or you may have a file cabinet containing valuable papers. You might want to protect these papers even if they were of no worth to anyone else. Not finding anything else of value, a vandal would think nothing of ripping and shredding your papers, and scattering them all over the floor. You can also use point protection to protect large power tools, valuable paintings, cabinets that contain drugs, plan files, and gun cases.

INSTALLING AN ALARM SYSTEM

Many home owners are quite handy, and have a fairly good working knowledge of tools, electricity, and wiring. If you are one of these, there is no reason why you cannot install your own home alarm system. The advantage of doing so is that by installing your own you will have the confidence to wire in all the windows and doors, instead of just one or two. Further, you will feel able to make any modifications or additions, gradually updating your alarm system from an elementary type to one that is quite sophisticated. Also, you will be able to make repairs should such repairs become necessary.

Having a system installed is more expensive since the cost of labor is quite often more than the cost of the materials. The advantage, of course, is the convenience of having someone do the job for you—someone who has all the necessary tools, knows from experience just where to mount the alarm, and can give you substantial advice on just which areas of entry should be protected.

INTRUSION DETECTION METHODS

Intrusion detection systems depend on a variety of electrical, electronic and mechanical devices. Alarm systems can make use of ultrasonic sound, infrared waves, photoelectric beam interruption, pressure sensitive devices, electromechanical relays, ambient light techniques, acoustics, etc. Basically, the idea is to be able to detect some change, whatever that change may be, in the normal home environment. The sound of a window breaking, or the noise made by a footstep, the action of a window opening—all of these, and more, are changes in a home environment that should not take place in the absence of its residents. It is these changes that are detected or noticed by the system that ultimately set off an alarm. In all instances, though, whatever method or technique is used, the purpose is to prevent the burglary.

WIRED VERSUS WIRELESS SYSTEMS

In some alarm systems, a change of state in a sensor, possibly caused by movement or pressure, closes a circuit that permits current in the control unit. Such sensors must be connected to the control device by wires, and so in a wired system, connecting the sensors to the control unit can become an installation problem. With a wired system, however, any number of windows and doors can be protected.

A wireless system eliminates the need for wiring, but the area that is covered by the intrusion unit is limited. It is true that more than one intrusion control can be used, but this raises the overall cost. Thus, there is no clear cut advantage of one system, wired versus wireless, over the other. It becomes a matter of personal preference and budget.

There are a number of ways of solving the alarm problem. You can buy a complete kit and do the installation yourself; you can have it done by a professional installer; or you can buy the unit parts, such as sensors and alarm, and design and install your own system. You can have an alarm system that will cover a complete house, or just a relatively small area. A small area protection system is sometimes used on the theory that a burglar will eventually move through every section of a house or apartment.

HOW A SIMPLE ALARM SYSTEM WORKS

Fig. 4-5 shows the circuit arrangement of a simple alarm system that is just about as easy to operate as an ordinary doorbell. The alarm device consists of a bell, or a buzzer or any other gadget that will make a noise. The alarm works when a switch is closed to allow a current of electricity through the circuit. The power

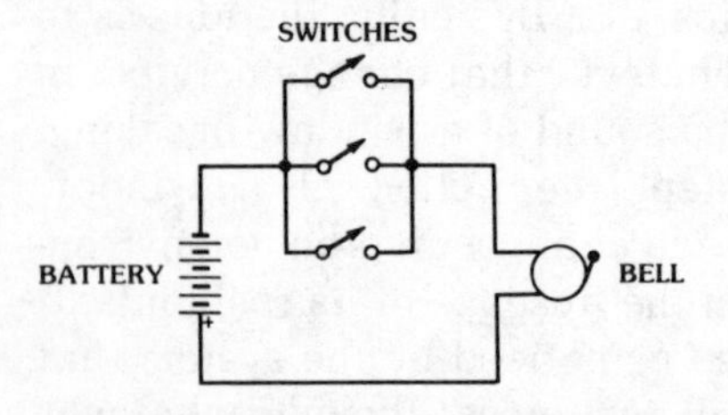

Simple alarm system circuit. When any one of the switches is closed, a current from the battery passes through the alarm, in this case, a bell. (Courtesy Hydrometals, Inc.)

Fig. 4-5.

source can be a number of dry cells or can be the nearest convenient ac outlet. When you buy the alarm, the manufacturer will supply you with information about just what source of power you should use. You can also get a complete alarm kit containing the switches, batteries, and alarm, plus full instructions on how to wire the system. It is better to do this than to buy the individual components, unless you have had some prior electrical experience. For example, one complete package is a door-and-window burglar-alarm system that fits all doors and that you can mount in minutes. It is completely self-contained, and is tamper-proof from the outside. It is fully transistorized and has a keyless electronic exit feature with an alarm panic button. The panic button is used to indicate an emergency, or to test the alarm system and the exit button. It is supplied with 25 feet of hookup wire, two keys, all the necessary mounting hardware, and operates on a 6-volt battery.

Another solid-state burglar alarm, closed-circuit system in kit form contains an 8-inch alarm bell, electrical switch lock, and a control-circuit panel housed in a heavy duty, weatherproof steel box that has a built-in tamper-proof circuit. Clipping wires or opening any protected point will cause the alarm to sound. The alarm will continue until it is reset by a key-proof lock switch. This kit includes four sets of magnetic door/window switches, two keys, warning decals, 100 feet of hook-up wire, insulated wiring nails and mounting hardware. It has an interior on/off switch for setting the alarm system from inside the house and a test/panic button to indicate an emergency or to test the alarm system. The alarm is powered by a 12-volt lantern battery.

CIRCUIT DIAGRAMS

When you buy a kit, you will be supplied with connecting in-

Fig. 4-6.

An alarm kit. (Courtesy Universal Security Systems, Inc.)

structions, probably including a pictorial diagram. A pictorial diagram, as its name implies, shows pictures or drawings of each of the components in the alarm system with lines, representing wires, interconnecting those components. There are also electrical and/or electronic symbols that are sometimes used instead. These symbols simplify the diagram, making the instructions easier for those who can read and understand such schematic diagrams. Sometimes a manufacturer of an alarm system will include both types of diagrams, pictorial and schematic.

Fig. 4-7 shows some representative alarm components and symbols used in circuit diagrams. The battery symbol shown in Fig. 4-7A is a series of long and short parallel lines. Plus and minus symbols (+ and −) are sometimes included, but can be omitted. The transformer in Fig. 4-7B is a step-down type that reduces ac line voltage to 6 or 12 volts ac for most alarm circuit usage. Fig. 4-7C shows three symbols for alarms: bell, lamp, and horn. The switches shown in Fig. 4-7D are single-pole, single-throw types. There are, however, numerous combinations of switches available. Any device that makes or breaks (opens or closes) a circuit can be called a switch. Fig. 4-7E shows the symbol for a relay. The abbreviation N.C. designates normally closed and N.O. designates normally open. The movable arm between the upper (N.C.) contact and the lower (N.O.) contact represents the

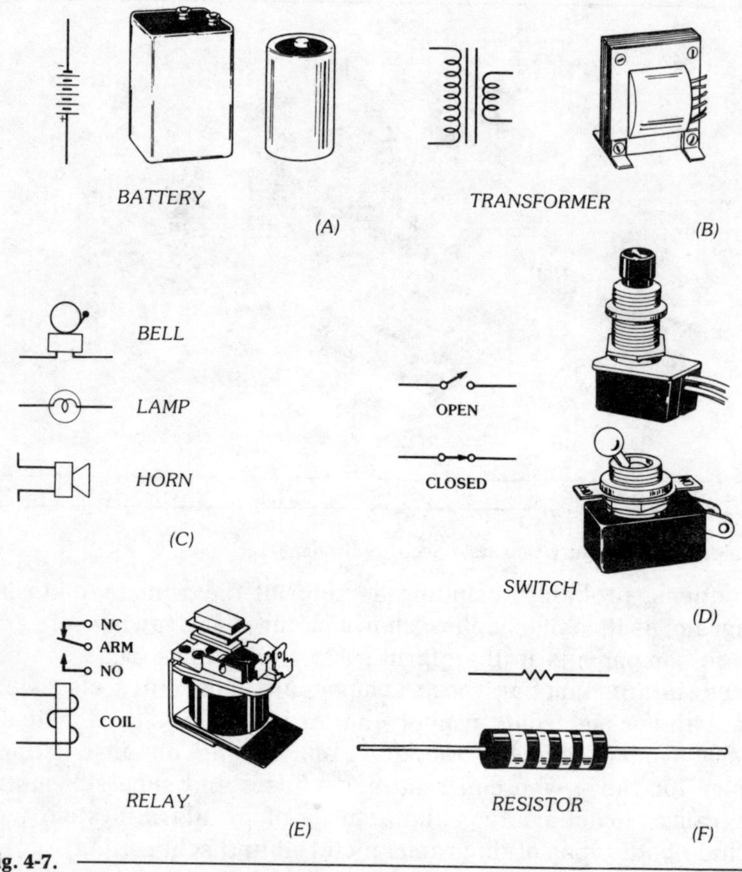

Fig. 4-7.

Circuit components and symbols used by manufacturers of alarm systems. (Courtesy Hydrometals, Inc.)

armature or blade switch. Fig. 4-7F shows a resistor, a device that limits the amount of current in a circuit.

COMBINED BURGLAR/FIRE ALARM

You can also get combined burglar-alarm and fire-alarm systems to give you double protection. And, while you are now thinking about security, it is a good idea to keep a flashlight in some convenient, easily reachable place near your bed. Should you think you have an intruder in your home, it will be easier and possibly faster to reach for the flashlight than for the light switch.

A burglar may also try to get to the power-line junction box in your basement to turn off the main power switch. The reason for doing so would be to deactivate any power-line-operated alarm devices.

Battery-operated alarm systems can be a nuisance in the sense that you must run periodic checks on the system to make sure the battery still supplies power. Even if a battery is not used, it will run down and in time will need to be replaced (if a dry type) or recharged (if a wet type). The advantages, though, of using a battery are that they seem to outweigh the inconveniences. A battery represents an independent power system. If your ac line voltage fails—and that can happen and has happened—you still have protection. Further, the alarm system cannot be deactivated by an alarm-conscious burglar who makes the fuse box his first stop.

Wired burglar-alarm systems work in a number of ways. Some are closed-circuit, others use an open circuit. Thus, in an open-circuit, any attempt by the burglar to clip the connecting wires will set off the alarm. In the closed-circuit type, closing a switch at various protected points will set off the alarm. The point is that the burglar simply does not know what to do.

When installing an alarm system, assuming you buy a complete package, you will receive a number of door and window switches. Sometimes you will have the option (depending on the manufacturer) of buying extra switches, to accommodate your personal needs. The switches are mounted on doors and windows in such a way that the action of opening them would close the switch. This would permit current to flow from the battery to the alarm. These switches could be supplemented by a mat-type switch, placed under a doormat or entrance rug.

PANIC BUTTON (EMERGENCY OR TEST BUTTON)

A panic button (Fig. 4-8) is used to trip or actuate an alarm system manually. Any number of switches can be incorporated into an alarm system. Usually, they are wired to the normally-open sensor terminals of the alarm control using ordinary 2-conductor wires such as No. 22 or No. 24 gauge alarm, intercom, or speaker wires. The test, or panic button, can be installed near the front door, in the kitchen or bedroom. Some people prefer putting the panic button on a night table adjacent to

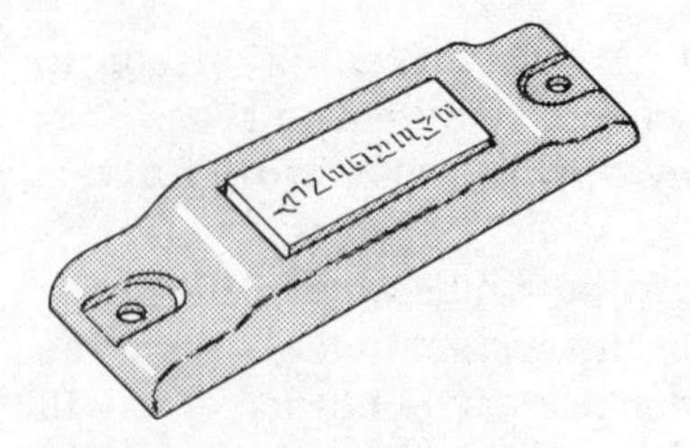

A panic button, also called an emergency or test button, is an ordinary normally-open switch.

Fig. 4-8.

the bed since they want the capability of sounding an alarm in case an intruder breaks in.

The panic button can be used to expand an existing alarm system. It supplies momentary contact and is used with normally-open circuit systems.

DEFEAT SWITCH

When an alarm system of this kind is triggered—that is, turned on—the alarm keeps sounding until a defeat switch is used. The defeat switch is part of the circuit and when operated interrupts the current to the alarm. When mounting the defeat switch, put it in some spot that isn't obvious to an intruder—that is, it should not be out in the open. With an alarm sounding, few burglars will remain on the premises to conduct a search. However, you and the various members of your family should know just where the defeat switch is, and should be able to reach it quickly and easily if the system turns on for any reason.

The location of the defeat switch and, as a matter of fact, your complete alarm protection system should not be the subject of discussion with neighbors or friends. The members of your family should know that the system exists and they should be able to locate the defeat switch in the dark, if necessary. You can be sure that for every alarm system invented, there will always be a burglar somewhere with an ingenious idea for overcoming it.

PULL TRAP

The basic elements of a pull trap are a switch and a length of wire. If the wire is pulled, it removes insulating material between the contact points of a switch, permitting the switch to close. The closed switch then allows an electrical current to activate an alarm, a light, or possibly some combination of the two.

While a pull trap can be simple and effective, and also low cost, it is an awkward and inconvenient alarm system. The trip wire can be accidentally pulled by a dog or by a child, or you may forget the trip wire is in place and set off the alarm. For the most part, pull traps have been replaced by more sophisticated systems.

Various types of pull traps are commercially available. One typical model consists of a rectangular section of insulating material placed between two spring-loaded bearings as shown in Fig. 4-9A. Fig. 4-9B is a pull trap used to protect a fence gate; Fig. 4-9C shows how a pull trap can be used to guard a garage door. With a wire trip cord (Fig. 4-9D), circuit connections can be made at the ends. In this illustration the pull trap is used to guard a driveway.

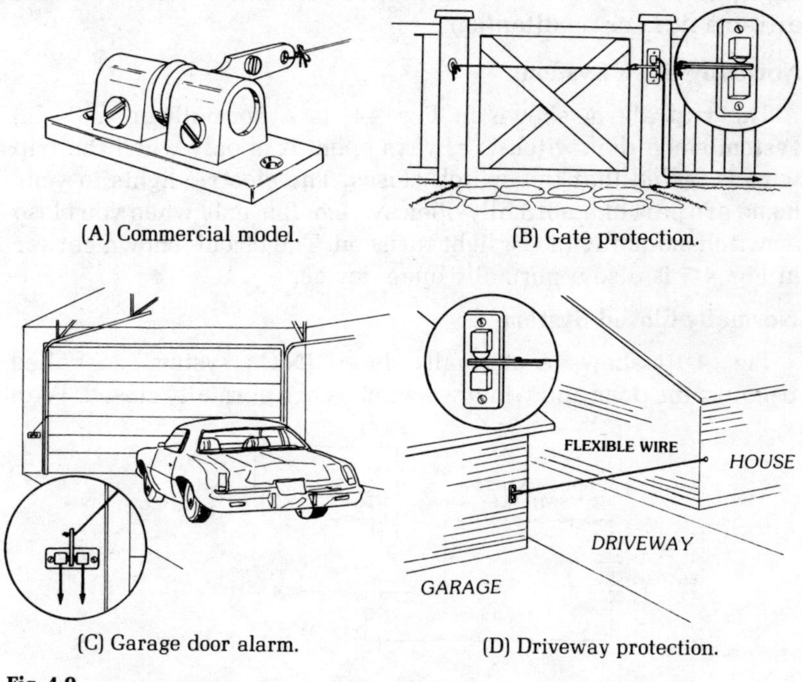

(A) Commercial model.

(B) Gate protection.

(C) Garage door alarm.

(D) Driveway protection.

Fig. 4-9.

Pull trap and various applications.

THE ALARM WARNING LABEL

When you buy a commercial alarm system for your home or apartment, you will probably receive a small self-stick label advising that your premises are electronically guarded. Use it. A burglar, noticing the label, may be deterred from trying to enter, and you will be saved the aggravation and expense of replacing broken window and door panes, jimmied locks, and damaged garage doors.

CIRCUIT SYSTEM ADVANTAGES AND DISADVANTAGES

The type of alarm system that uses open switches is identified either as an open-circuit alarm or normally-open (N.O.) system. While it is simple, inexpensive, and effective, it has a number of limitations that reduce its reliability. For example, if one of the wires leading to one of the switches is cut or broken at any point, the alarm will not work. Confronted by such a system, a burglar might be able to cut or snap a wire so quickly that the alarm attracts little or no attention.

Normally-Open System

The trip alarm shown in Fig. 4-9 is a normally-open (N.O.) system since the switch is always open. It is only when the trip wire is pulled that the switch closes. The electric lights in your home are part of a normally-open system. It is only when you close a switch manually that a light turns on. The circuit shown earlier in Fig. 4-5 is also a normally open device.

Normally-Closed System

Fig. 4-10 shows a normally-closed (N.C.) system, so called because the door and window switches are normally closed. With

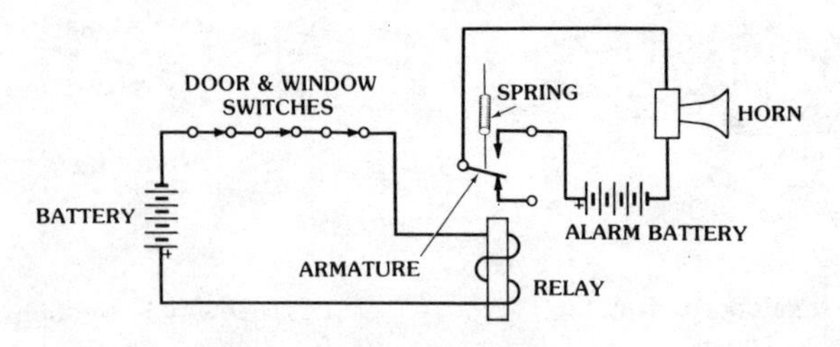

Fig. 4-10.

Normally-closed switch circuit for alarm system.

the arrangement shown in the drawing, current flows from the battery, through the door and window switches, and through the coil marked relay. The current makes the relay coil into an electromagnet. The magnetism of the relay holds the armature pulled down as shown. Thus there is no current from the alarm battery to the horn.

Now assume that any one of the door or window switches opens. This could be caused by a burglar opening a window or the door. The current from the battery at the left is interrupted and so the relay ceases to be an electromagnet. The spring pulls the armature, a sort of blade switch, to the upper contact. Now the circuit to the alarm (a horn, in this case) is closed, and since current from the alarm battery passes through the horn, it is activated and sounds an alarm.

This arrangement is more difficult to defeat than an open-circuit arrangement. It is generally called a closed-circuit system, but because there is a supervisory current through the sensors (the switches) and relay at all times, it may be called a *supervised* system. Since a failure of any part of the sensor circuit (all of the switches) will cause the system to sound an alarm, it can be identified as a *fail-safe* design.

While far superior to the simple open-circuit alarm system, the basic closed-circuit system has a disadvantage of its own. If an open sensor switch is closed, once opened, the system will return to its pre-alarm condition. Thus, an intruder who has accidentally tripped a door or window switch can silence the alarm simply by restoring, or by putting a short across the opened switch. The shorting device can be a short length of copper wire, or it may be a length of wire ending in a pair of spring clips so the burglar need not hold it, but can clip it into position. Once this is done, the burglar can proceed about his business, fully aware now of the type of alarm system confronting him. However, there is a circuit design, as shown in the illustration, which can once again make the burglar's work as perilous as it should be.

Nonresetting Alarm

Fig. 4-11 shows an alarm circuit in which a resistor, a current-limiting device, is used across a reset switch. The resistor limits the current to a value needed to hold the relay closed, but not to close it. If a burglar opens any of the door or window switches or breaks the window foil, the relay will de-energize, causing the alarm to ring. The alarm will continue to ring, even if the burglar resets the door or window switch he has opened. This is a little more complex setup than the alarm system described earlier since it prevents the burglar from silencing the alarm.

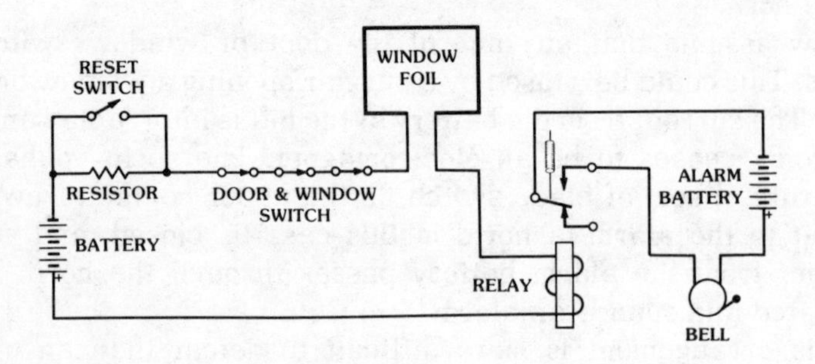

Fig. 4-11.

A burglar cannot turn off this complex alarm by operating the reset switch.

The new circuit is quite similar to the basic closed-alarm system. Several new items have been incorporated, however. One of these is window foil (described in Chapter 8) while the others consist of a reset switch and a resistor. A resistor, shown symbolically by a series of lines that look like sawteeth, is a part that limits or opposes the flow of an electrical current. When the resistor is in the circuit, the amount of current is comparatively small. When the reset switch is closed, it effectively shorts or removes the resistor from the circuit, and so the amount of current is increased.

Electromagnetic relays have an interesting characteristic. They require more current to close their armatures than to hold their armatures in a closed state. In other words, it takes quite a bit of current passing through the armature coil to make it into a magnet that is sufficiently strong to attract and pull over the armature. But once this is done, once the armature is indeed pulled over, the amount of current needed to hold it in that position is relatively small.

Now, by adding a resistor of the proper value to the circuit, as shown in the illustration, we can limit the supervisory sensor current to the value needed to hold the relay closed, but not to close it, if opened. To close the relay, should it be opened, we short the resistor momentarily by means of the reset switch.

The burglar now has a much tougher circuit to overcome. If he accidentally opens the sensor circuit, by opening a door or any sensor-switch-protected window, the armature of the relay will be pulled away toward the bell-ringing circuit and the alarm will sound. And the alarm will continue to sound, even if he restores a tripped switch or shorts a broken lead. In fact, the only way he can silence the alarm is to either locate and disable the alarm circuit itself, or to locate and operate the reset switch.

The odds are against him, however. The alarm circuit itself can be contained within a strong tamper-proof box, and even if the burglar is able to locate it, and do so quickly, he must work under conditions he always tries to avoid, knowing that the alarm must soon bring someone to investigate. He may also look for the reset switch, but in a well-designed installation, the reset switch itself would be key-operated. The burglar is now faced with a switch that requires a key, and again, picking the lock for the reset switch, or smashing it, or trying to overcome it would take time, and with the alarm ringing that is just what the burglar does not have. The reset switch can not only be locked, but it can be positioned within a tamper-proof cabinet. Faced with these difficult-to-solve alternatives, the intruder generally will disappear to a less noisy location.

MAGNETIC SWITCHES

There are many different types of switches, but no matter how they are made, the job they have is the same and that is to open or close a circuit so an alarm can be sounded. In addition to conventional switches, you can buy plunger or pushbutton switches, pull traps, switch mats, window foil, and key switches. There is still another type of switch sensor now popular with intrusion-alarm systems and that is the magnetic switch, as shown in Fig. 4-12.

A magnetic switch contains two basic parts—the switch and an actuating magnet. (Courtesy Hydrometals, Inc.)

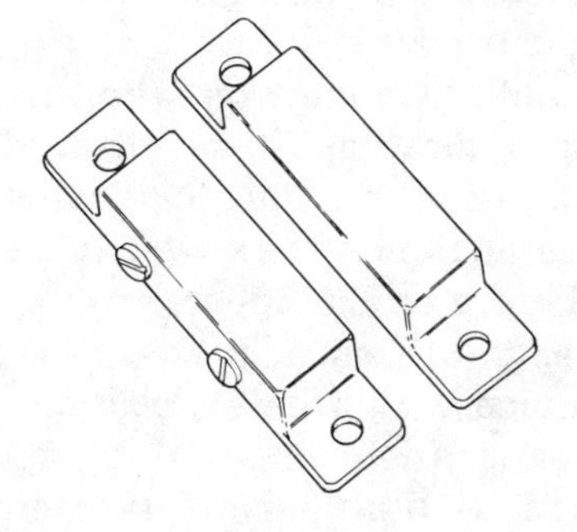

Fig. 4-12.

How a Magnetic Switch Works

The switch (Fig. 4-13) consists of two or more reedlike metallic structures that carry the electrical contacts and are magnetically sensitive. Under normal conditions, they are held apart by spring tension and thus the two metal strips inside the switch enclosure do not touch. In the lower drawing, a magnet is brought near the switch, pulling the flexible upper strip down to the fixed lower metal strip. As soon as the contact ends of the two strips touch, the switch is closed. If the magnet is moved away, spring tension opens the contacts.

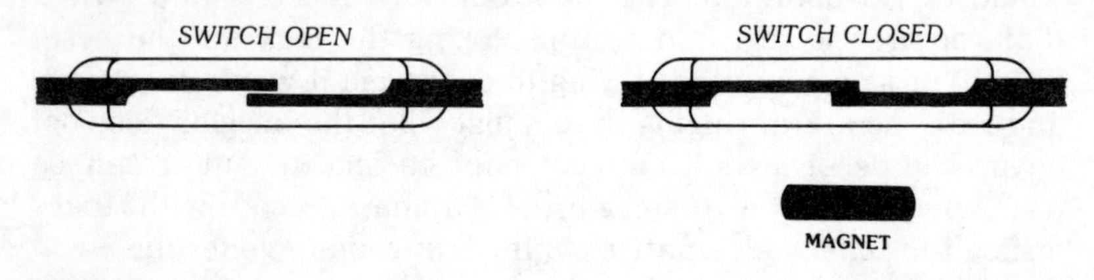

Fig. 4-13.

The reed switch closes when magnet is brought near it. (Courtesy Hydrometals, Inc.)

Magnetic Switch Installation

In practice, the magnetic switch is normally mounted on a window or door frame, as shown in Fig. 4-14. Wired into a closed-circuit system, the switch will open and trigger an alarm whenever the magnet is moved. Since the magnet is mounted directly on the window frame, the window cannot be opened without moving it. But when this happens, the switch will open and trigger an alarm. Two magnets are provided on the window to maintain protection whether the window is closed or partially open for ventilation. Either of the two different magnet/switch mounting positions for a door (Fig. 4-15) is satisfactory for most installations. The switch/magnet can be mounted anywhere along the top or the upper portion of the side where it will be out of reach of young fingers. Two switches are shown in Fig. 4-15 to indicate choices of location, however, only one switch/magnet combination is needed for the door alarm. Wiring for the switch can be tacked into place along the door jamb with insulated wiring nails.

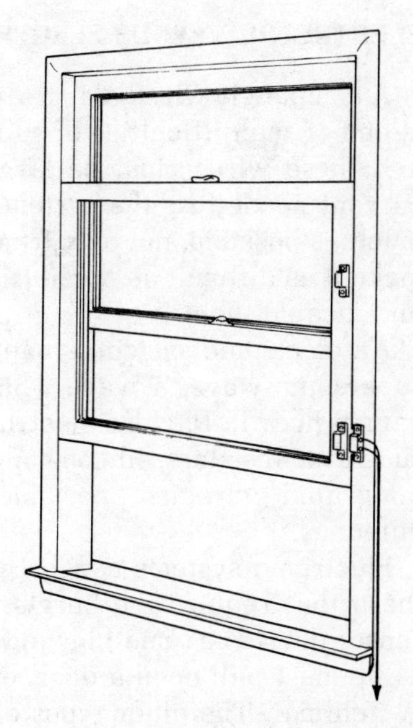

Basic arrangement of sensors on a window.

Fig. 4-14.

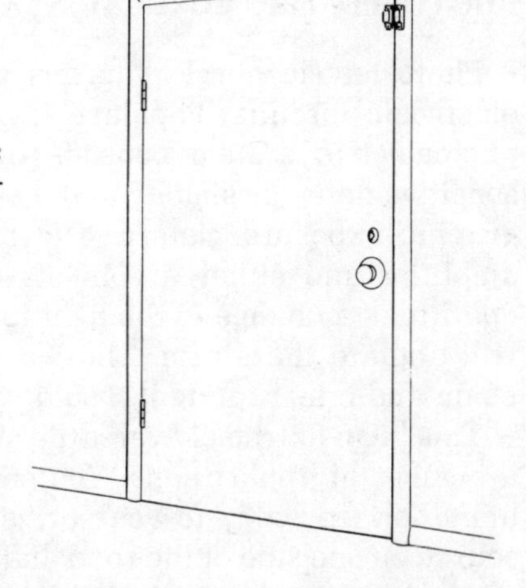

Two possible positions for magnetic switches on a door. (Courtesy Hydrometals, Inc.)

Fig. 4-15.

ELECTRONIC VERSUS ELECTRICAL SYSTEMS

A completely wired alarm system is an electrical type. And that is one of the difficulties of such a system. It requires long wires and these wires must be large enough to carry the amount of current needed by the system. The wires must be concealed as much as possible, not only from an aesthetic point of view, but to make it as difficult as possible for the burglar to be able to locate and disable them.

An electronic system is one that depends on a radio wave or supersonic wave, a wave whose pitch is so high human beings cannot hear it. Modern electronic systems use space-age devices such as transistors, silicon-controlled rectifiers, infrared sensors, integrated circuits, ceramic transducers, and microwave antennas.

Electronic systems can be grouped into two basic categories: those that require the burglar to make physical contact with a sensor device or something to which the sensor is attached— that is, he must pull open a door, open or break a window, step on a switch mat. The other types of electronic alarms are those that require no physical contact for they can detect an intruder when he interrupts a beam of light, just enters a certain area, makes a sound, or just moves about.

PHOTOELECTRIC BURGLAR ALARMS

Photoelectric burglar alarms were among the first to utilize electronic circuits. Popularly called an electric eye, the basic photoelectric system consists of a light source and a light-sensitive detector similar to the elements used in ordinary photographic exposure meters. The detector can be coupled to an amplifier and either a solid-state or electromagnetic relay. In operation, a change in the light falling on the detector (or sensor) will actuate the alarm. Thus, a burglar stepping into the light beam and interrupting it would cause the alarm to sound.

This is an extremely versatile alarm system and can be used in a number of applications. You can use it, for example, to detect unauthorized entry to your private driveway by putting a light source on one side of the road and the sensor on the opposite side, as shown in Fig. 4-16. Here the light beam will be interrupted by an approaching vehicle. A similar arrangement can be used to guard a gate, door, or window, or even a whole line of doors or windows, as along one side of a hallway or building.

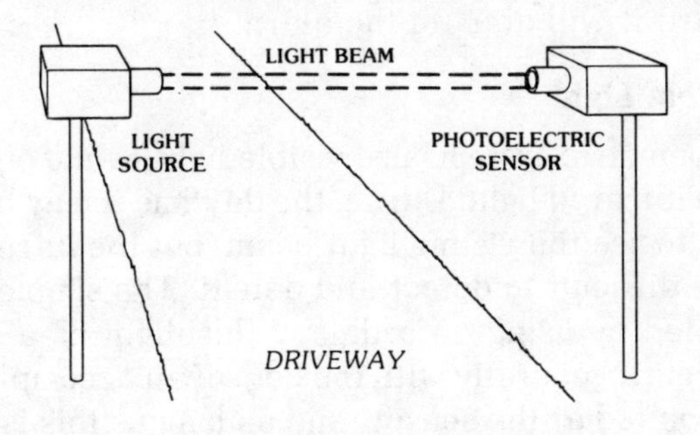

Fig. 4-16.

Protection of driveway by photoelectric method.

Two or more photoelectric systems, together with mirrors, can be used to set up a "light fence" around an enclosure or a home affording complete perimeter protection.

Photoelectric systems can also be adapted to supply area and point protection. Fig. 4-17 shows a technique in which mirrors are used to crisscross a light beam back and forth across an entire area. If the light beam is interrupted at any point, the alarm will go off. You can get point protection by directing a light beam at a mirror attached to the protected object, such as the door of a

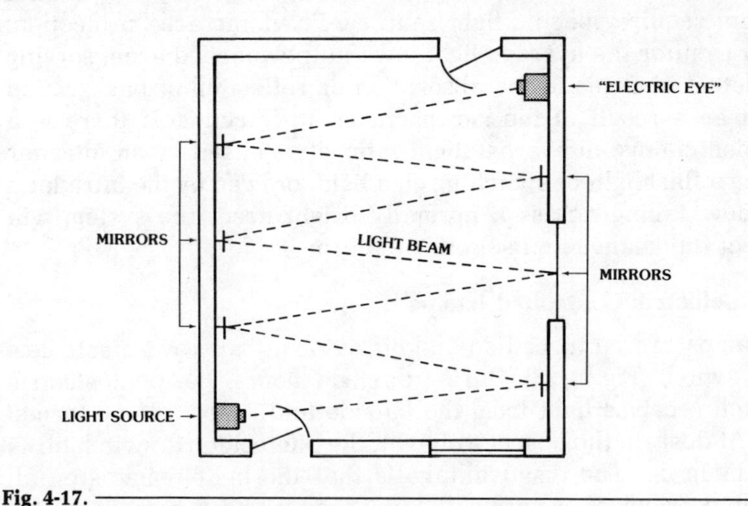

Fig. 4-17.

Photoelectric technique for area protection.

safe, and using the photoelectric sensor to pick up the reflected light. Anyone standing in front of the object or moving his hand directly toward it will actuate the alarm.

Visible/Invisible Light

Some photoelectric systems use visible light, while others work with invisible infrared light. During the daytime it may be difficult for a burglar to see the visible light beam, but the infrared beam is much more difficult to detect and defeat. The simpler systems can be defeated by using an ordinary flashlight as a substitute light. The burglar, generally with the help of an accomplice, keeps a flashlight pointed at the sensor, and as long as this is done, the alarm will not go off. A simple metal stand to hold the flashlight firmly in position can be used, so that the burglar is certain that the alarm will not be tripped. To defeat the burglar, one technique is to use an interrupted light beam with the electronic circuit designed to respond only to such a source. The light pulses can be made to occur so swiftly that to the burglar the beam is a steady one. However, any attempt to use a substitute steady light source, such as a flashlight, including an infrared flashlight, will activate the alarm. Even if the burglar is aware that the beam is an interrupted type, there is no way of knowing if the light pulses are all uniform, or how often they occur.

A few manufacturers offer photoelectric security alarms that do not require specific light sources. Providing area protection, they monitor the average light level in the guarded area, serving to detect changes in the absorption or reflection of background light as a result of the movement of an intruder. If there is a sudden change in average light intensity, caused by an intruder using a flashlight or switching on a light, or even by the intruder's shadow falling across a normally bright area, the system will detect the change and set off the alarm.

Photoelectric Controlled Lamps

Lamps can be turned on and off automatically by a photoelectric switch (Fig. 4-18). During daylight hours, the photoelectric switch receives light from the outside and keeps a lamp turned off. At dusk, in the absence of light, the photoelectric switch turns the lamp on. The disadvantage is that the light remains on all night, not a usual or normal lighting sequence in a home. A better arrangement is to have several timers that turn lights on and off

Fig. 4-18.

A lamp can be controlled by a photoelectric switch.

in various parts of the house or apartment, but that turn off completely at some more usual hour, possibly around midnight.

You can also get a combined light/photoelectric timer. The unit plugs directly into any convenient ac outlet and, as long as there is adequate light in the room, remains turned off. As soon as the ambient light falls below a certain level, the light turns on automatically.

Like other security devices, these units have advantages and disadvantages. They are low cost and require no attention. Once they are plugged into an ac outlet, they require no manual switching. However, they do remain turned on all night since they are plugged into an outlet and most outlets are along a baseboard where the light level is usually low, they generally turn on long before dusk and often do not turn off until late in the morning, not a normal or usual lighting situation in a home.

The light source in the combined light/photoelectric timer is a 7½-watt night light. The bulb is easily replaced and the unit can be regarded as an automatic night light.

THE CAPACITY ALARM

A pair of metal plates, separated by air, or some other insulating material such as mica, is known as a *capacitor*. When the capacitor is connected to a source of voltage, such as a battery, the capacitor will become electrically charged. The amount of charge depends on the area of the metal plates, how close the plates are to each other, and the material between the plates. If, for example, you bring your fingers close to the capacitor plates, the electrical charge on those plates will change. There is no need to touch the plates; the proximity of your fingers is enough.

This phenomenon is used in another type of electronic alarm system, which senses changes in the electrical charge on a metal plate as a person approaches. Known as capacity alarms, these systems are often used for point protection. There are some sophisticated versions available that use long wires instead of plates—the basic idea of a charge on a plate remains the same—and it can supply excellent perimeter protection.

SOUND AND VIBRATION ALARMS

Another type of electronic alarm system is one that responds to a sound. The simplest uses a sensitive microphone as a sensor, with the microphone connected to a high-gain audio amplifier similar to those used in hi-fi systems. Others use special vibration sensors similar to the "tilt" detectors in pinball machines. When adjusted for maximum sensitivity, the vibration detector can pick up a soft padded footstep across a room of fair size.

The trouble with sound and vibration alarm systems is that they can produce frequent false alarms. Thus, the passage of a heavy truck can cause a vibration-type alarm to go off.

ULTRASONIC ALARMS

The ultrasonic alarm is one of the most effective of the modern electronic intrusion-detection systems. The ultrasonic system uses air vibrations similar to those emitted by "silent" dog whistles, that produce sound outside the range of human hearing (Fig. 4-19).

In operation, an ultrasonic alarm system radiates a very high frequency or ultrasonic signal into the protected area with the help of a special type of loudspeaker. It then detects the echoes that bounce back to it. The pitch of the echo is compared electronically with the original ultrasonic wave sent out by the loudspeaker. Under normal conditions, both sounds, the original and the echo, will be identical. Should there be a moving object within the area, such as an intruder, the echoes will be changed. This difference will be detected and will serve as a control signal to set off the alarm (Fig. 4-20).

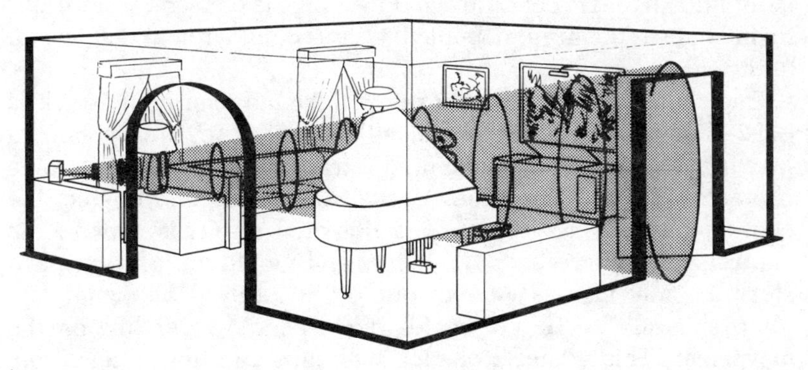

Fig. 4-19.

Ultrasonic waves are invisible and inaudible.

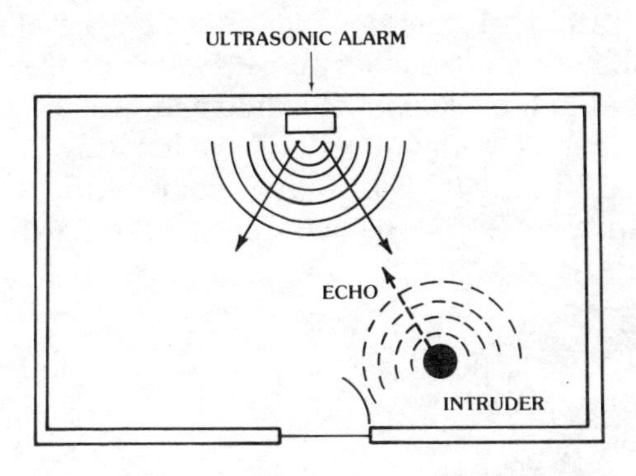

Fig. 4-20.

Ultrasonic generator radiates inaudible waves, covering selected area. Intruder steps into wave area, causing ultrasonic echo to return to sensor, actuating alarm.

Capable of extreme sensitivity and ideal for area protection, ultrasonic alarm systems suffer from one limitation: They will respond to any movement, even air movement within their range. As a result, pets and children must be kept out of the area. Fans and blasts of air from an air conditioner can also set off the alarm.

The operation of an ultrasonic space alarm is based on a law of physics known as the Doppler principle. This principle states that if there is movement within an area of sound waves, the movement will change these sound waves at a given rate. This change and rate can be sampled and detected. If the change is such that it would be caused by human movement, the control unit is set into an alarm condition. If the change is caused by any other turbulence in the area, it would be filtered out without causing an alarm.

One sound system fills an area with sound energy at 20.2 kHz (20.2 thousand cycles per second), a frequency above human hearing. This is done by mounting a transmitter in the area to be protected. In the same area, a receiver is mounted to detect the continuous 20.2-kHz signal. The detected signal is sent to the control unit, where both frequency and amplitude are sampled. Human movement will change the 20.2-kHz signal by approximately 40 Hz (40 cycles per second), depending on the movement. This 40-hertz change will cause an alarm condition. Any other movement, such as swaying curtains, window rattling, etc. will change the 20.2-kHz signal appreciably above or below the 40-hertz rate, and thus have no effect on the alarm system.

Fig. 4-21 shows an ultrasonic alarm system concealed in a book that can be put in a bookcase, at random on a shelf, or on a table. The purpose is to keep the intruder from locating the control unit and possibly disabling it. The waves generated by the ultrasonic generator can protect up to 300 square feet, with a teardrop

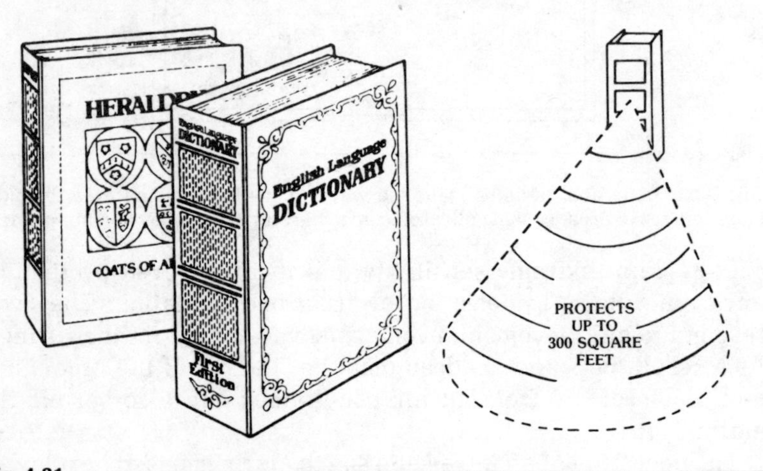

Fig. 4-21.
Book conceals ultrasonic alarm system.

pattern, depending on environmental conditions. Any significant motion occurring within this pattern will cause the alarm to respond.

When motion is first detected, a lamp or other appliance connected to the unit will turn on immediately. Seconds later, the internal alarm and/or an optional external alarm horn will sound. The appliance and the alarm will remain on as long as the intruder is present and for approximately one minute after motion has stopped. Then the alarm and the appliance will turn off and the unit will automatically reset itself, supplying continuous coverage.

Another unit is an ultrasonic alarm that blankets an area of up to 600 square feet (67 square meters). It sounds an 85 decibel dual-tone alarm when an intruder enters the protected room.

The unit, an area protection alarm for enclosed rooms, has an approximate 60-second preset exit delay and a 30-second preset entry delay. It can also be used to drive an external public address horn. The control (Fig. 4-22) is set on a three digit code. There are many different codes so it is most unlikely someone you know will get one with the same code.

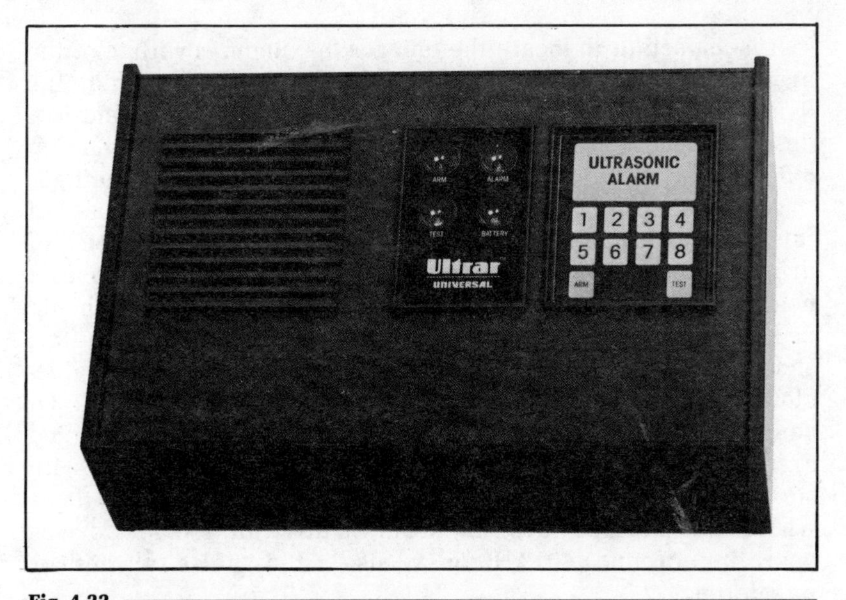

Fig. 4-22.
Control box for ultrasonic alarm. (Courtesy Universal Security Instruments, Inc.)

As shown in Fig. 4-23, the ultrasonic signal fills the rooms with a cone-shaped pattern that expands vertically and horizontally as you move away from the alarm. A sensitivity control on the rear panel of the control unit can be used to adjust pattern coverage to the size of the area to be protected.

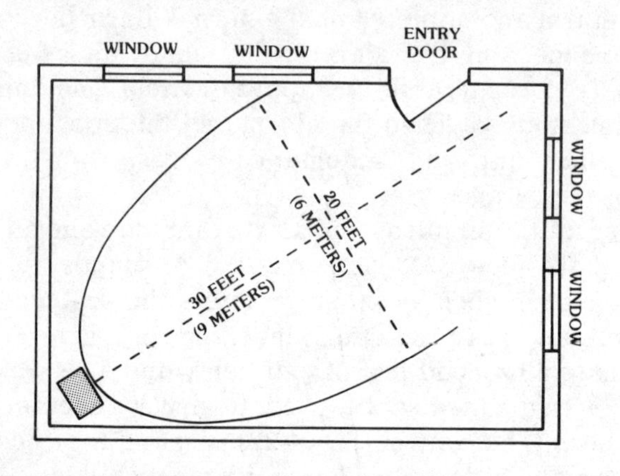

Fig. 4-23.

Area covered by ultrasonic alarm when positioned in corner of room. (Courtesy Universal Security Instruments, Inc.)

It is important to locate the unit for maximum coverage of the area to be protected. Usually the corner of a room, such as the living or living/dining area in a smaller house or apartment is a desirable location. Alternatively, it can be placed at the end of a long hallway to guard the access to all doorways along the hall.

With systems of this kind there can be a number of causes of false alarms, such as the movement of a pet, or the fluttering of drapes in the protected area due to air currents from a heating/cooling system. While you are away, pets at home should be kept in another room. Drapes near a heat duct, likely to flutter, should be weighted or clipped, as should any loose-fitting tablecloths or slipcovers on furniture that might be affected. Of course, any movement of family members through the protected area will also set off the alarm. For this reason, families with small children may not be able to use this security device at times when the children are at home. Other ultrasonic devices operating around 40 kHz may also cause false alarms or interference.

If an intruder enters while the unit is armed, the alarm will begin sounding about 30 seconds after the protected area has been disturbed, even if the intruder has already passed through the protected area and is elsewhere. The alarm will continue approximately 3 minutes, after which it will reset itself automatically.

FREQUENCY

A wave transmitted by a control unit can be ultrasonic, infrared, or ultra-high frequency (uhf). The difference among these waves is just a matter of the frequency. An ultrasonic wave could be some frequency somewhat higher than the range of human hearing, possibly 30 kHz (30,000 cycles per second). An infrared wave is one whose frequency is just outside the visible spectrum at its red end. The frequency of an ultra-high frequency transmission could possibly be around 200 MHz (200 megahertz or 200,000,000 cycles per second).

Each of these waves has advantages and disadvantages. An ultrasonic wave could be blocked by a wall, or much of its energy could be absorbed by house furnishings such as drapes and carpets. A wave in the uhf region could pass through a brick wall, but might not have as much reflectivity as an ultrasonic wave.

WIRELESS ALARM SYSTEM

One example of a wireless alarm system consists of a small ultra-high frequency radio transmitter at each protected entry point and a programmable alarm receiver with an internal sounding device.

When a transmitter is triggered by moving the associated magnet, disturbing a wired sensor, or pushing the panic button, a radio signal is sent to the central alarm receiver or control. The receiver then senses the channel setting of the transmitter, observes a delay period if any has been set, and sounds the appropriate signal and/or external alarm.

The ultra-high frequency radio signals have a range of up to 300 feet (92 meters) allowing an entire house and even an unattached garage or shed to be protected with the wireless alarm system. The internal alarm sound is loud and will be audible throughout most homes. The external speaker can be located almost anywhere outside the house (Fig. 4-24).

Remote alarm-horn speaker. (Courtesy Universal Security Instruments, Inc.)

Fig. 4-24.

Fig. 4-25 shows the placement of the magnet and transmitter on a standard door. Opening the door moves the magnet, turning on the transmitter. The transmitter can also be mounted away from the door, connected by a pair of wires to a magnetic switch. A similar arrangement can be used on a double hung window, or sliding windows, or doors. (See Fig. 4-26).

The control box of the uhf wireless system has a keyboard, as shown in Fig. 4-27. As its name implies, the keyboard permits complete control of the wireless system, with each function determined by a separate key. Thus, the timing of the alarm can be decided by depressing any one of nine pushbuttons, with the timing ranging from as short as 5 seconds to as long as 45 seconds. The entire system can be tested, or the battery. The keyboard is equipped with a silent key.

COMBINED PROTECTION

An alarm system is not a substitute for locks nor are locks an alternative for alarms. Both are needed if you want maximum security. Whether you should have both depends on individual requirements. If you live in an apartment that is in a very low crime-rate area, and if the lobby is security-guard protected, and if you have had a pair of front-door locks installed by a master locksmith, then you might consider your protection adequate. If you have a home with substantial amounts of silverware, jewelry, and furs, then what is sufficient for an apartment may not be enough for you. The purpose of a lock is to keep a burglar out; the object of an alarm system is to stop the burglary and/or to cause the apprehension of the burglar.

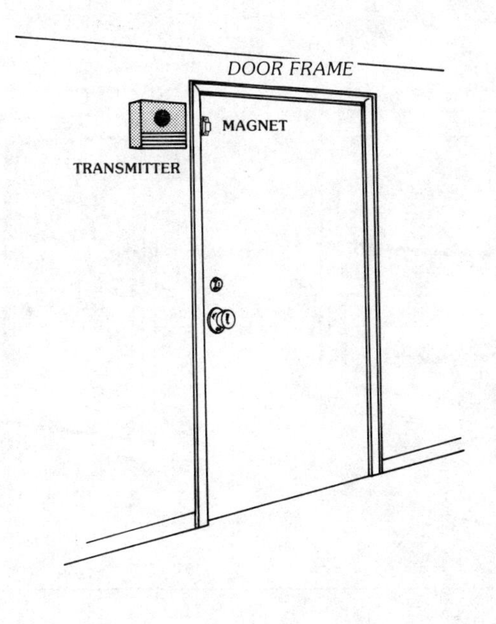

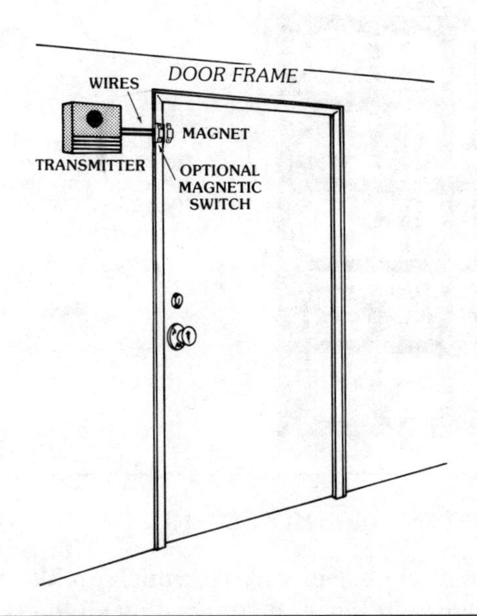

Fig. 4-25.

Door installations of ultra-high frequency transmitters. (Courtesy Universal Security Instruments, Inc.)

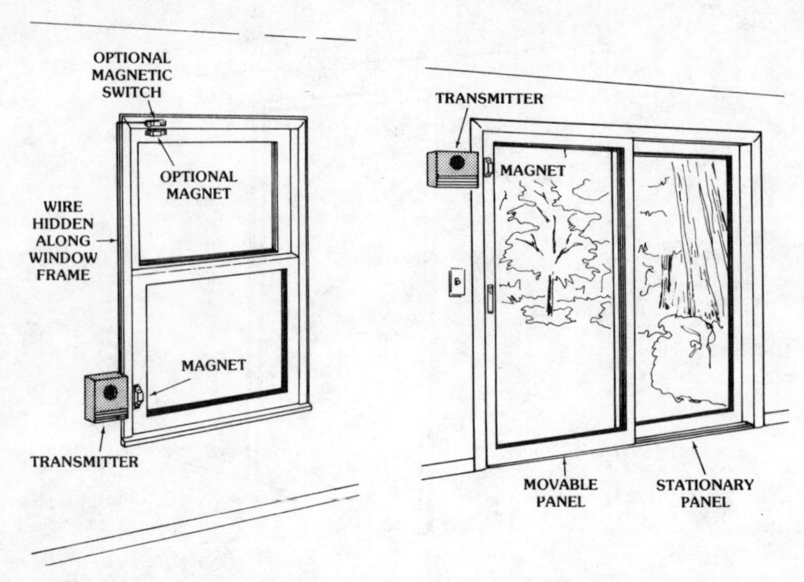

Fig. 4-26.

Typical transmitter installations. (Courtesy Universal Security Instruments, Inc.)

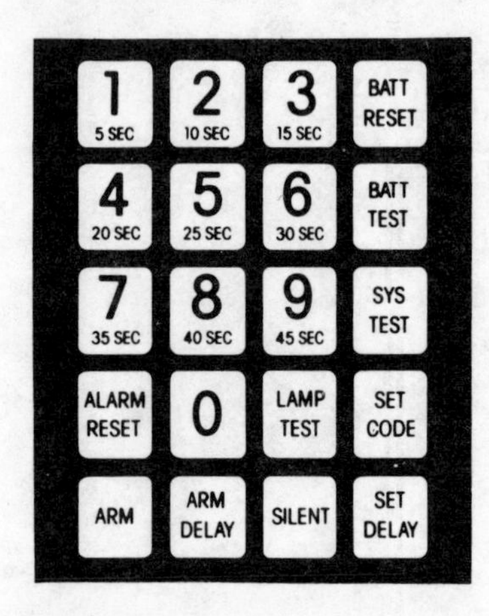

Wireless uhf alarm system keyboard. (Courtesy Universal Security Instruments, Inc.)

Fig. 4-27.

ALARM SYSTEM ARRANGEMENTS

There are a number of ways in which an alarm system can be set up, as shown in Fig. 4-28, depending on the type of protection wanted. A pair of magnetic switches (A) can be used on doors and windows. A pull trap (B) can guard a path, road, or driveway; (C)

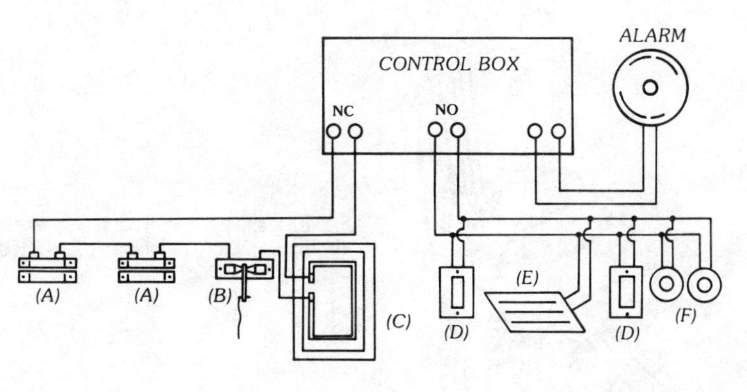

Fig. 4-28.

Alarm systems can be arranged in many different ways, depending on protection wanted. (Courtesy Hydrometals, Inc.)

shows a normally-closed window-foil installation and (D) is a normally-open test/panic button. A mat-type switch (E) can be placed right before the door on the inside or at the foot of a stairway. A pair of fire sensors is shown at (F). For a home it is unlikely that more than one type of alarm arrangement would be used. It all depends on what entry areas are to be protected. A large house could possibly have one alarm system for the downstairs, another for upstairs. A business might have a number of alarm systems, possibly all of the same type, so as to cover as great an area as possible.

WHERE TO PUT SENSOR SWITCHES

Most often sensor switches are installed on the door or doorjamb or on one or more windows. But these are precisly the areas that burglars expect you to install switches and so an experienced burglar may work out some technique for defeating them. Security in the home is an unending battle between those who want to break and enter and the homeowner who wants to keep burglars out.

There are many places in the home where switches can be installed for the purpose of turning on an alarm to warn of intruders. A flexible switch in ribbon form shown in Fig. 4-29 can be used under the padding of a chair (Fig. 4-30). If the burglar sits in the chair to rest, or sets something in the chair, the switch can activate the alarm. This same type of thin switch stitched or

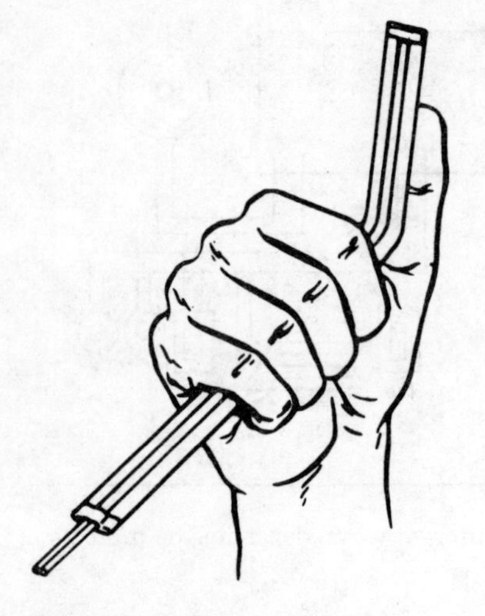

Flexible switch has numerous security applications. (Courtesy Tapeswitch Corp. of America)

Fig. 4-29.

Switch can be imbedded in any cushioned furniture. (Courtesy Tapeswitch Corp. of America)

Fig. 4-30.

cemented into the draperies, as shown in Fig. 4-31, will detect an intruder's entry whenever the seams are flexed. A ribbon switch can also be put under a mattress on a bed (Fig. 4-32), or it can be placed across a doorknob as shown in Fig. 4-33. A simple turning of the knob in either direction will actuate the switch. When placed behind a valuable painting, this type of switch can sound an alarm if the picture is moved (Fig. 4-34). Such switches can be mounted in or behind a drawer that contains valuables (Fig. 4-35), or placed on a safe as shown in Fig. 4-36 so that any movement will cause an alarm to sound.

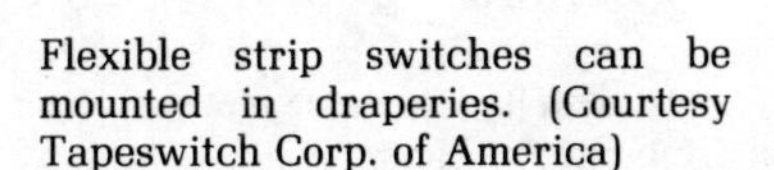

Flexible strip switches can be mounted in draperies. (Courtesy Tapeswitch Corp. of America)

Fig. 4-31.

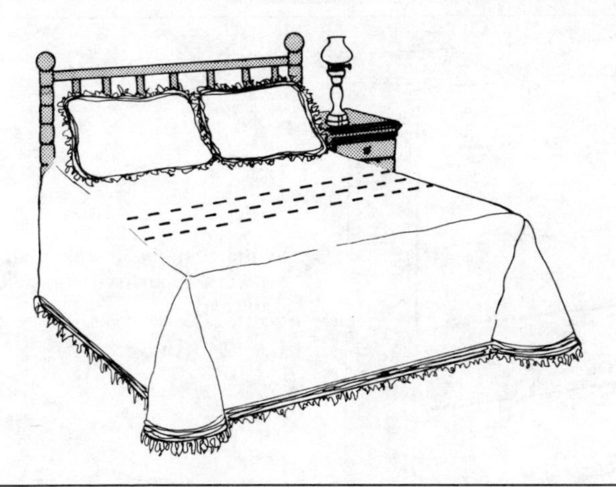

Fig. 4-32.

Even a bed can be wired to signal an alarm. (Courtesy Tapeswitch Corp. of America)

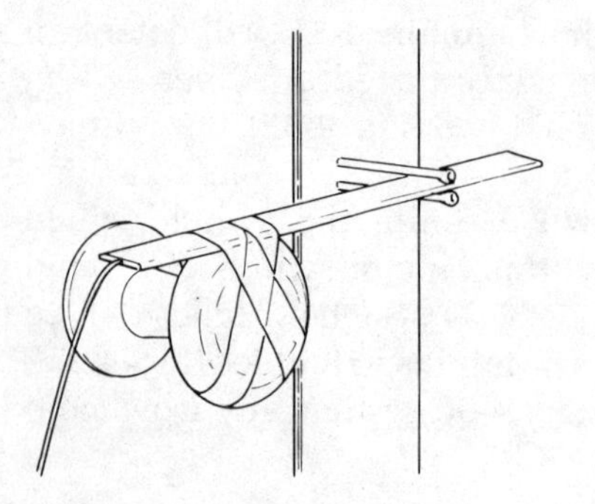

Turn the doorknob in either direction and the switch actuates. (Courtesy Tapeswitch Corp. of America)

Fig. 4-33.

Flexible switch or strip-type switch can be mounted behind a painting. (Courtesy Tapeswitch Corp. of America)

Fig. 4-34.

Switch can be installed in or behind drawers. (Courtesy Tapeswitch Corp. of America)

Fig. 4-35.

If wheel-mounted safe is moved out of position, switch actuates, setting off alarm. (Courtesy Tapeswitch Corp. of America)

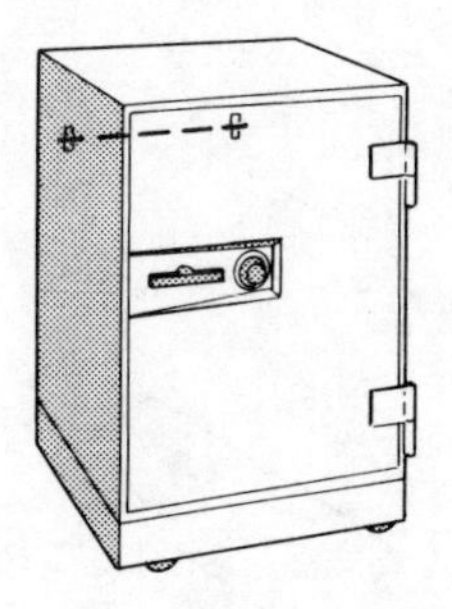

Fig. 4-36.

ALARM TIMING

Generally, if an alarm rings in a house, a burglar will try to make his getaway in as few seconds as possible. However, since the alarm has been set off, it will continue to blast until someone stops it. Since the alarm has achieved its purpose, there is no point in having it blast continuously. Some alarm systems do just that, but others have an on-off cycle. Some alarm systems have automatic timing, with a certain amount of time on, then a certain amount of time off, enough to give you the time to reach the defeat switch. With some alarms, the timing is built in and you cannot change it. With others, you can set the alarm timing any way you want.

With an alarm system in the house it would be advisable to notify your nearest friendly neighbor just what the system is and show where the defeat switch is located. It would also be a good idea to take your local police into your confidence.

THE PROBLEM OF SWITCHES

Switches present two basic problems. The first is visibility, the second is the method of connecting the switch to the control unit. One technique is to use pressure sensitive switches and to conceal them beneath room rugs, scatter rugs, or step carpeting. This is an advantage since the switches are hidden and it is quite likely the burglar will step on one of the rugs. For greater assurance, a number of pressure sensitive switches can be used. (Fig. 4-37). The difficulty, of course, is in connecting the switches to the control unit.

To eliminate the need for wiring under rugs, a miniature transmitter can be connected directly to the switch. In Fig. 4-38, a pair

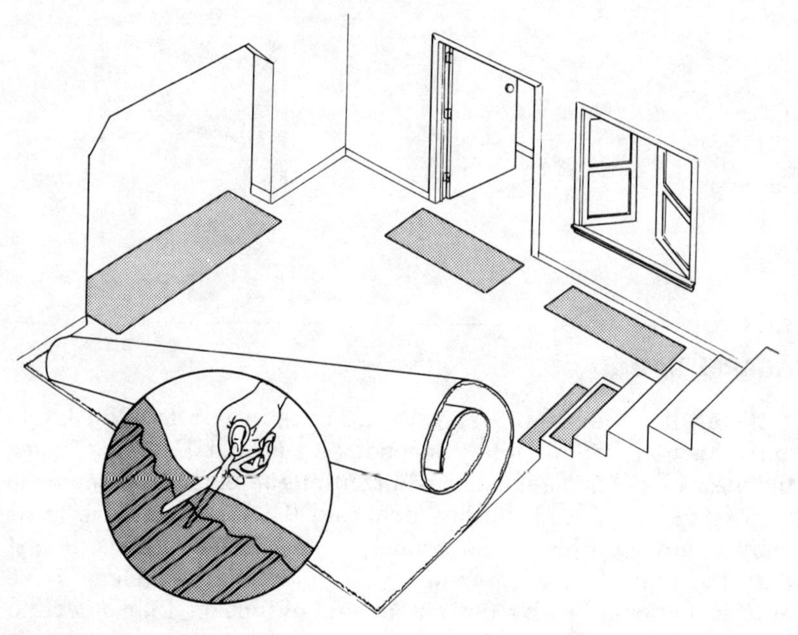

Fig. 4-37.

Mat switches are pressure-sensitive switches and can be positioned beneath scatter rugs, room-size rugs, or stair-step carpeting. (Courtesy Hydrometals, Inc.)

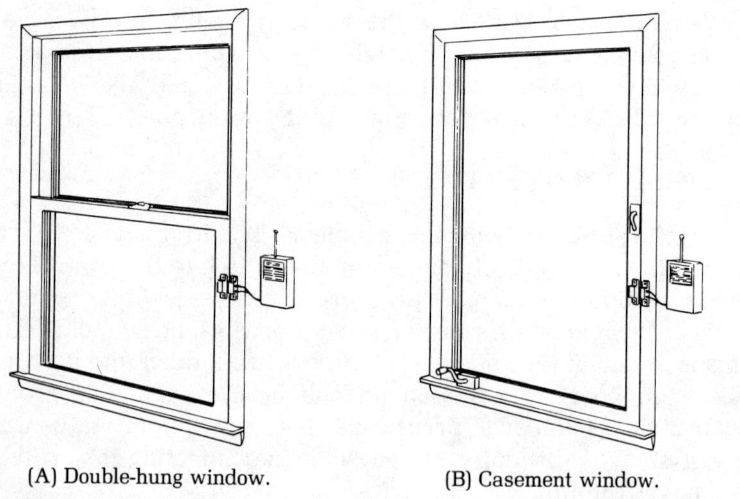

(A) Double-hung window.

(B) Casement window.

Fig. 4-38.

Transmitter installation on double-hung and casement windows. (Courtesy Hydrometals, Inc.)

of magnetic switches are put on a double hung window and on a casement window. When the windows are disturbed the adjacent transmitter is set into action and transmits a signal to a receiver in a control box, which, in turn, activates the alarm.

The setup in Fig. 4-38 does eliminate wiring, but it has a number of disadvantages. One of these is cost, for a separate transmitter is needed for each area of entry. Wiring, while more of a nuisance, is much less expensive. Another problem with the installation of Fig. 4-38 is that the thief, in making his entry, might notice the transmitter and take immediate steps to disable it. This might or might not disable the alarm, depending on its design. If the alarm is the type that continues to sound once it has been started, disabling the transmitter would do the thief no good.

Flexible switches can be used on overhang, casement, or basement windows, as in Fig. 4-39. The same type of switch can be used on a door (Fig. 4-40). And, if you need to make electrical connections to alarm devices mounted in a hinged access enclosure you can use highly flexible two-wire cables (Fig. 4-41). The cable, however, is not a switch, and is simply used when it is necessary to run a connecting pair of wires to a control unit.

Fig. 4-39.

Flexible switch can be used for overhang, casement, or basement windows. (Courtesy Tapeswitch Corp. of America)

POLICE ALARM

With our current crime statistics and the rate they are mounting, you have every reason to be concerned for your safety and you should do everything within your power to ensure it. But

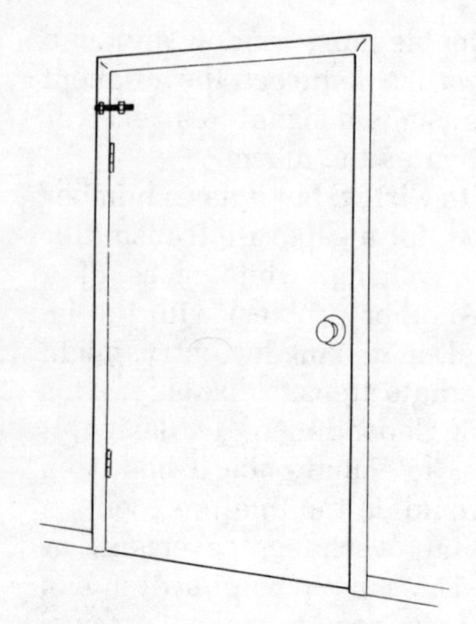

Opening the door can operate a switch to set off an alarm. (Courtesy Tape-switch Corp. of America)

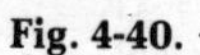

Fig. 4-40.

Fig. 4-41.

Door cords are highly flexible two-wire cables. (Courtesy Hydrometals, Inc.)

not everyone cares to have an alarm system in his home. An alarm can frighten the occupant of a home about as much as a burglar. Some alarms are so loud and so insistent that they can even scare home owners who know about them and who can anticipate the sound. And some people find an alarm a traumatic shock. There is little doubt that elderly people and those with heart conditions may suffer. The solution is the use of a telephone alarm. The telephone alarm can be programmed to respond to any emergency condition. It is an automatic telephone reporting device that communicates emergency information by prerecorded tape to the proper authorities. The minute an intruder tampers with any protected opening in your home, the telephone alarm instantly telephones the police, automatically calls for help, and gives the exact location of the emergency. The telephone alarm is a compact telephone communicator that operates on regular house current. It will automatically switch over to a built-in battery standby unit in the event of a power failure, thus ensuring continuous operation under all power conditions. For each forced entry, the telephone alarm will make three or more separate phone calls. However, it does not interfere with the regular use of your telephone, as it is directly connected into special equipment supplied to you by your local telephone company.

WHERE TO POSITION INTRUDER ALARMS

If you decide to install an intruder alarm of the wireless or ultrasonic type, then you must come to some decision about the area you want to cover. For example, the pattern of the alarm may not cover the entire floor or room. However, in most cases, a single, properly situated intruder alarm is capable of safeguarding you effectively. If you have selected an ultrasonic type of alarm, consider these possibilities as "burglary-trap zones."

1. Hallways or corridors connecting one section of your home with another. If a burglar is to get something out of your home, he will need to move from one room to the next, unless he is already thoroughly familiar with the layout and has already made a prior decision about what he wants to steal.
2. Stairways.
3. Areas with exterior openings that offer possible entrance for burglars.

Be certain that the burglar-trap zone is within the wave pattern of your alarm system. Don't select areas where small children or pets can enter the trap zone and cause false alarms.

SOME DON'TS OF INTRUDER ALARM INSTALLATIONS

Intruder alarms often come equipped with a sensitivity control. Do not advance this control—that is, make the unit extremely sensitive. If you do, you will find you will have a number of false alarms. Don't use the intruder alarm in rooms with open windows. Don't permit children or pets to enter an area when the intruder alarm is switched to the alarm position. Don't use an intruder alarm outdoors.

Some intruder alarm systems do a double job. They not only sound an alarm, but turn on a light when the protected area is entered.

THE ADD-ON SYSTEM

When homes or apartments are located in high-crime areas, it may be difficult or expensive to get property insurance. In some cases insurance becomes available only after the resident agrees to install some type of alarm system. Some alarm systems are being designed for "add-on" capabilities, so you can begin with a starter system and then gradually add more and more protection. The advantage here is that you will get some basic protection, and you will be able to buy a system that is within your budget. And then you can add on more sensor switches, or have greater area coverage when you can afford to do so. And if you are building a new home, you should know that some building contractors are now including alarm systems as an option. One of the advantages of doing so during construction is that it is much easier to install concealed wiring before walls are erected.

DOOR AND WINDOW ALARMS

There are numerous inexpensive alarm devices available but such gadgets are often not worth the room they require. Not only do they have extremely limited use, but they may instill a false sense of security, depriving the house owner or apartment dweller of adequate protection. One such gimmick is the simple door or window alarm, usually completely self-contained in a thin plastic box and powered by Penlite cells. Such cells aren't really designed for this purpose, have a short life, and require frequent replacement.

Devices of this kind shouldn't really masquerade under the heading of alarms, for they are really annunciators and are simply a modified form of door buzzer, now doing double duty as an alleged alarm system.

These bargain basement alarms sound their buzzer when a door or window is opened, but stop just as soon as either one is closed again. They do serve some purpose, for example, as a bedroom-door alarm or in a small shop to alert the proprietor that someone has entered a rear or front door. You can also use them on your house doors to alert you when children come or go. As far as security is concerned, you can put one or more on your bedroom windows to wake you should a burglar try to make entry that way. The weak sound produced by the buzzer shouldn't be considered as having deterrent value.

THE ALARM PROBLEM

Installing an alarm system and then hoping for the best isn't enough. If the alarm is indoors it may not be heard if the alarm isn't loud enough and if all windows and doors are closed, as they probably are. A better arrangement is to have an inside and outside alarm, a step that is easier to take with a house than an apartment. But even with an apartment the possibility exists that the owner of the building might permit the use of an outdoor alarm.

Many alarm signaling devices go unnoticed because of their steady, monotonous sound. The best arrangement is a siren sound that changes its pitch, something a bell-ringing alarm is unable to do. And a cycled alarm with a varying pitch is still better. This means the alarm will have an on-off sequence. You can add an outboard module that will vary the pitch and cycle of an alarm but this adds to the cost. That is why some alarm systems cost more than others—they supply more. It may be more expensive to buy an alarm system and then add outboard components to it than to purchase the best system the first time.

Horn Ratings

Horns are rated in terms of their voltage and current requirements by the amount of input power, and by their sound output in decibels (dB). Current to a horn can range from a fraction of an ampere to several amperes. Horn voltage is either 6 volts or 12 volts dc, but some are also designed to work from the 115-volt ac power line.

The dc-power input to a horn is the product of the voltage and current going to it—that is, the amount of current in amperes multiplied by the amount of voltage in volts. If a horn takes 1 ampere of current at 6 volts, then its dc-power input is $1 \times 6 = 6$

watts. The greater the power input to a horn, the louder it will sound.

When two or more horns are used, they are connected in parallel. The total power used is the sum of the individual powers of the horns. If you connect two 6-watt horns together, their total power operating requirement will be 12 watts. Since the voltage of the battery connected to the horns does not change, the greater the number of horns, the larger the amount of current flowing to them. Thus, if you have two horns connected to a 6-volt battery, with each requiring 1 ampere current, the total current will be 2 amperes. The power consumed is $2 \times 6 = 12$ watts. If the horns are identical, each will have an operating power of 6 watts.

The wires connecting the horns to the horn battery must be capable of carrying the required current. The amount of current will depend on the number of horns you use. If you use four horns, with each calling for a 0.5 ampere current, then $4 \times 0.5 = 2$ amperes. The connecting wire must be capable of carrying this amount of current. The current-carrying capability of wire depends on its cross-sectional area, with the length of the wire having some effect on the total current the wire can carry. The thicker the wire, the more current it can handle. The longer the wire, the greater the resistance or opposition of the wire to the flow of current. Before buying wire for an alarm system, then, you should know how much current each alarm will require and the total length of wiring required. The total length is twice the distance from the battery to the alarm since you must have one wire from the battery to the alarm and another from the alarm to the battery. If the battery is 20 feet from the alarm, you will need a minimum of 40 feet of wire. Make allowance for the fact that the wire will usually not be connected in a straight line between the battery and the alarm.

The Alarm Battery

The voltage of an alarm battery is fixed, and can be 6 volts, 12 volts, or some other value. The horn rating must match the voltage rating of the battery. Thus, you would use a 6-volt horn with a 6-volt battery, a 12-volt horn with a 12-volt battery.

The current-delivering ability of a battery is rated in ampere hours, and is an indication of the number of amperes a battery can deliver over a period of time. A 20-ampere-hour battery, for example, can theoretically supply 1 ampere for 20 hours or 2 amperes for 10 hours, or 4 amperes for 5 hours. The ampere-hour rating isn't precise since horns, bells, and sirens aren't usually operated over long periods of time. The larger the ampere-hour rating of a battery, the heavier it will usually be, and the more it will cost.

Batteries don't last forever, but are self-consuming, and so their current-delivering capacity constantly decreases. This is a very slow process, so a good plan is to try the test button about once a week just to make sure the alarm system is operative. If the alarm goes on but the tone seems to become noticeably weaker as the sound is prolonged, the usual trouble is a weak battery. If you are using a rechargeable type of battery, you can bring it up to operating condition by connecting it to a charger. It's a good idea to put a battery on charge for a short time, about a half-hour or so, at least once a week. It is possible to determine the state of charge of a battery with a voltmeter, but to do so the alarm must be connected and working. If the battery voltage has decreased by 20 percent it is time for a recharge. Thus, if a 6-volt battery measures somewhat less than 5 volts, you should recharge it. To make the test use a dc voltmeter, with the plus lead of the meter (usually a red-coded lead) connected to the plus terminal of the battery and the minus lead (usually coded black) to the minus terminal of the battery. You must turn on the system—that is, the alarm, siren, or bell must be operating—during the voltage measurement. You can do this by pushing the test/panic button of the system. A voltage measurement without the horn blasting will be meaningless since you will probably find that under such test conditions the battery will indicate full voltage. Of course, if the system uses two or more sirens, possibly one inside and the other out, both should be on during the voltage test.

And Finally . . .

1. Some of the more elaborate alarm systems seem to be expensive. And while they represent one of the best kinds of insurance, remember that most people have to be persuaded into getting insurance.

2. While the owner of a home may have no installation problems, it is quite another matter for the apartment dweller. Many leases require that the tenant may not do any wiring in the apartment, or if an alarm system is installed, it becomes part of the landlord's property and must remain when the tenant moves.

In many cases the tenant is afraid to inform the landlord he wants an alarm system since this could lead to a demand for a rent increase.

3. Many people have an "it can't happen to me" attitude and will not consider alarm systems for that reason.

4. Quite a few people have a rather odd notion about our police departments. Their attitude is that they pay taxes, that it is the function of the police to protect them, and that if they are robbed, it must be the fault of the police. Their common complaint is: "You never see a policeman around when you need one." Such an attitude is not only childish and immature, but is completely self-defeating. The only way—the only possible way—is to acknowledge and realize that personal security is your job, and that it is a job you must do with the help and cooperation of the police.

Chapter 5

How To Protect Your Apartment

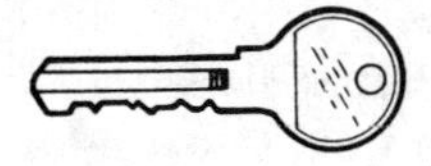

Some information on how to protect a house or an apartment was supplied in Chapter 2. Quite often security measures for a house or an apartment are the same and so the two types of residences can be grouped. However, an apartment dweller is not only as vulnerable as someone who lives in a house, but, in some respects, is more so. A general tendency of apartment residents is that they "do not want to become involved," and so if they see a stranger at the entrance to an adjacent apartment, they pay little or no attention on the theory that they are minding their own business. This is a form of rationalization and really means: "I don't care who gets robbed, so long as I am not the victim."

Those who live in apartments may not have as many windows and doors to secure, but often one of the windows, either front or rear, exits onto a fire escape. Further, certain areas of the apartment house—the laundry, the elevators, and inside garage—have all been the scenes of robberies. And leases often contain restrictive clauses that do not permit the tenant to make lock changes or to use alarm systems. But security for an apartment is essential since, on the average, every 60 seconds another apartment is burglarized. You can get insurance, but this will often not cover all of your loss and you may lose possessions with personal significance that cannot be replaced. Also, in some high-crime areas you may be able to get only limited coverage or none at all. In crime areas, it may not be possible to insure furs and jewelry.

MOVING INTO A NEW APARTMENT?

If you move into a new apartment, the first thing you should do is to change the front door lock. Make this a condition of your lease *before* you sign. Some cities now have an ordinance that makes it illegal not to change the lock.

LOCKS FOR APARTMENTS

The best kind of lock is one that is equipped with a 1-inch deadbolt. And get a lock whose manufacturer emphasizes that it is pick-resistant. Finally, the lock should have a double cylinder. A double-cylinder lock is one that requires a key for entry to the apartment, but you must still use a key to get out.

There is a good reason for having a double-cylinder lock. Some apartment doors have thin wooden panels and it is no great effort for a burglar to push these through, reach in, and turn the lock open. But this can't be done if the lock needs an *inside* key. Still another reason is that the burglar, unable to get in the front door, may try and succeed in getting in through a window. But now there is the problem of getting heavy objects out. With the front door locked from the inside, he will be unable to move heavy objects—color tv, stereo equipment, etc.—out of the window. This doesn't mean you won't get "burgled," but it does mean that the burglar's options have become more limited.

A good pick-resistant double-cylinder lock is a step in the right direction, but you are still faced with two problems. With a double-cylinder lock, you effectively lock yourself into your apartment. In the event of an emergency, you may not have time to start a search for your keys. For this type of lock, it may be a good idea to keep a spare key, placed somewhere near the door but out of sight and where it is unlikely to come to the attention of a burglar. Keep this as your "panic" key, your emergency exit key.

Another problem with the cylinder lock is that a burglar equipped with a cylinder puller can remove the cylinder in short order. To prevent this you can use a cylinder guard. This is a metal plate that covers the cylinder. It has a circular cutout in it so you can still insert your key into the cylinder, and it makes it impossible for the burglar to use a cylinder puller.

Brace Lock

Still another security accessory is the *brace lock*. This consists of a steel rod that extends from your door lock with the other end fitting into a metal socket in the floor. Once the burglar defeats your door lock. The next step will be to push the door open. The brace lock, should keep the burglar from doing so, if it is properly installed. If it is loose, the burglar may be able to move the door back and forth enough to pull the lock out of its floor socket.

Lock Chain

Make sure you have a lock-type chain on your door. This will enable you to open the door wide enough to allow packages to be slipped in without inviting strangers into your apartment. Even if the person who delivers a package to you isn't a burglar, he can pass along information about the layout and your personal possessions.

Window Latch

The center-swivel latches (Fig. 5-1) you will find on your windows supply as little protection in an apartment as they do in a home. If there is a space between the upper and lower window a burglar can slide a wire through and open this sliding-type latch.

Sash lock and strike for double-hung windows. (Courtesy Stanley Hardware, Div. of The Stanley Works)

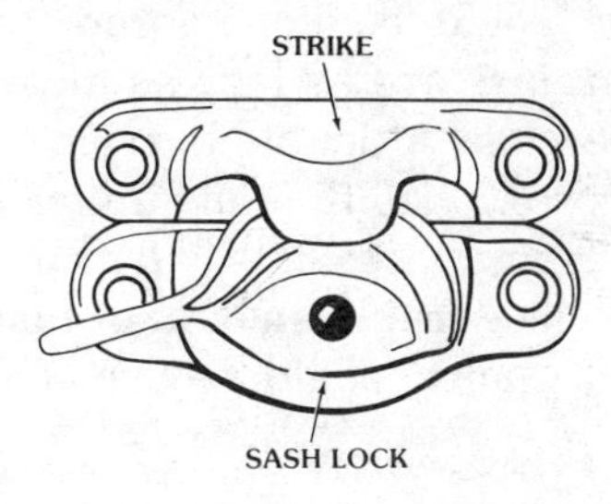

Fig. 5-1.

If not, he can tape one of the windowpanes, break it with little noise, and then reach in and open the latch. The best type to use on a window is a lock that works with a key (Fig.5-2). The lock mounts at either top corner of the movable window and fastens it to the window jamb. When activated, it locks the movable window in the closed position, or it can be locked in the partially open position for ventilation. One-way screws can be used for security.

This is costlier than the swivel latch but it is the only real apartment-window security you can have, other than a switch that turns on an alarm, or lights, or both.

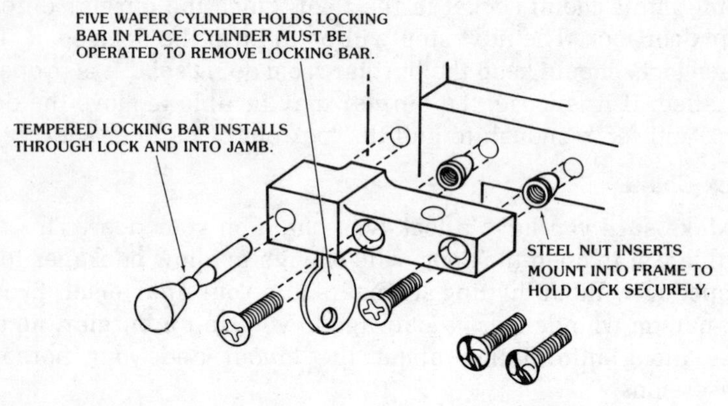

Fig. 5-2.

Keyed window lock for wooden and single-hung aluminum windows. (Courtesy Taylor Lock Co.)

Supplementary Locks

Still another burglary deterrent is to have two locks on a front door instead of the usual one. These can be two different types, one could be jimmy-proof and the other highly pick-resistant. Burglars are as lazy as anyone else and the thought of trying to overcome a double barrier can act as a psychological deterrent. There is also the time factor involved. Most people get in and out of their apartments rather quickly. The sight of someone spending a considerable amount of time in front of a door, possibly making some noise, could arouse concern in some neighbors.

DOORS

Unlike a house, entrance to an apartment is limited. Not all apartments are the same, of course, and some have a large number of windows, and possibly a rear entrance or a patio door.

Door Jamb

In older apartment buildings, you may find the wooden door jamb suffering from old age. Since the door jamb holds part of your lock. it may give way to just several energetic pushes, taking the lock with it. Examine the door jamb. The door should fit right against it.

To prevent the use of a shim, the jamb may have an outside section of wood or metal trim. The trim may form a frame around the door or may simply be along the side that has the strike. This protection isn't always used. For such installations you can replace the strike with a type that does not permit the use of a shim, as shown in Fig. 5-3.

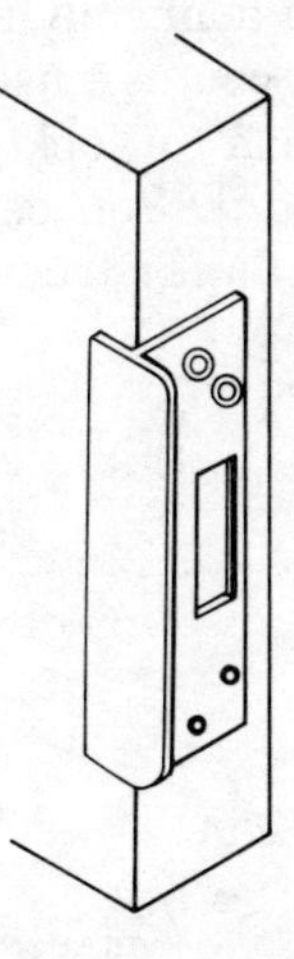

This strike does not permit the use of a shim.

Fig. 5-3.

The Strike

The strike is usually a rectangular section of metal fastened in the door frame, receiving the bolt (or bolts) of the lock. Sometimes the metal strike is removed and the only thing that holds the bolt is a cutout in the wood of the frame. This is useless. If the strike is missing, replace it with a reinforced type. None of the screws holding the strike in position should be loose; if they are, the screw threads have either rusted or the door jamb has become defective.

Terrace Door

Just because your apartment is a few stories above the street does not mean your terrace door need not be locked. Some burglars are quite athletic and have been known to go from one apartment to another by climbing up, terrace by terrace, trying all doors until they find one that is open. Treat a terrace door just as you would your front door. Use a quality double-cylinder lock, and make sure there is a sturdy deadbolt. Keep an extra key hidden on the terrace. If you lock yourself out, a terrace can become a very uncomfortable place. And, if you have a window facing the terrace, it too deserves a window lock or window lock plus alarm system.

The terrace door may be equipped with a lock, but this gives you only two options. Either the door is open or it is closed. If you want the door only partially open, but you want it locked in that position, you can use the sliding door fastener shown in Fig. 5-4. Since it isn't a locking type, remember that a thief can reach in and undo it. You can minimize this possibility by mounting the fastener as far from the open area of the window as possible.

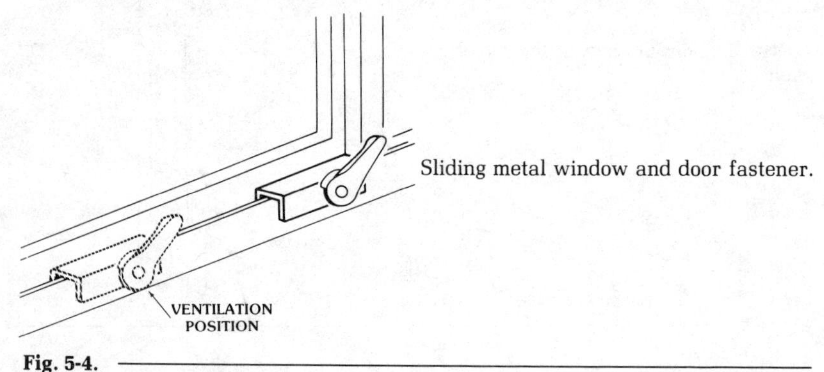

Sliding metal window and door fastener.

Fig. 5-4.

Peephole

Your apartment door should have a peephole (Fig. 5-5). Quite an old ploy is to have someone come to your door claiming they have a package for you and that you must sign for it. The peephole won't let you know if the messenger has a legitimate purpose or not, but at least you can see if the messenger has a uniform, or if there is more than one person. Ask them for the name of their company, and then, without opening the door, get the phone number from the phone book or the operator and verify that you do indeed have a package scheduled for delivery to you. As a further safety check, ask for the name and a short description of the messenger.

If the messenger is indeed someone trying to get into your apartment the person will try to pressure you into opening the door, insisting that there are many deliveries to make and that he doesn't have time. Let the person wait.

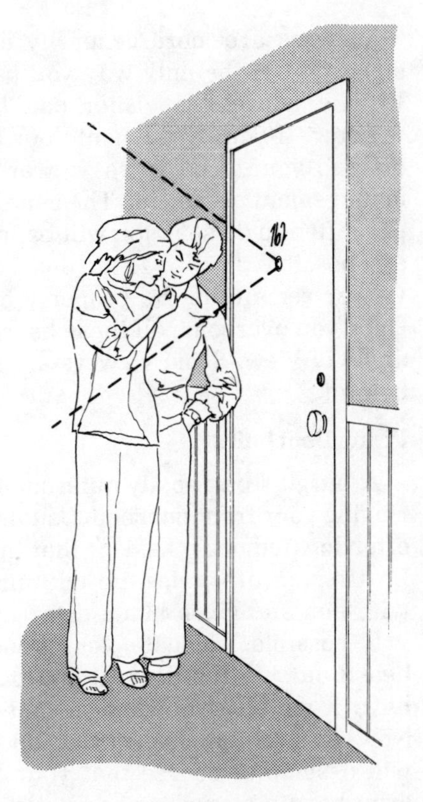

One-way door viewer permits you to see caller without opening the door.

Fig. 5-5.

Nor should you open your door to anyone else, ranging from door-to-door salesmen to service people such as plumbers and electricians claiming to have been sent to you by the management of your building. If you do have a problem in your apartment—leaky faucet, a window that is stuck, etc.—make sure that you know just who will be sent to your apartment by the building superintendent: that is, the name of the service company and the expected time. Even with these precautions, it is always better to have at least one other person with you at the time the repairs are made.

The best arrangement is a combination of viewer and locked door chain. When someone comes to your door and rings the bell, it is almost an instinctive reaction to swing the door open. Don't. Instead look through your viewer, and if you are satisfied that the message or package for you is legitimate, then unlock your slide chain fastener.

Viewers are more generally used on apartment house doors since that is the only way you can see who is outside the door. With a house, the visitor can be observed through any front window. If your apartment door doesn't have a one-way viewer, discuss your need for a viewer with the landlord or building-management personnel. The usual arrangement is that you will be permitted to do so if you will bear the expense, if you will pay for any possible damage to the door, and if you will agree to have the viewer remain as the property of the landlord when you move. Once you overcome all these barriers, get a viewer having a wide angle of view. Some viewers are very restricted and aren't very helpful.

Front Doorbell

A burglar can easily determine if you are home or not just by ringing your front doorbell. Usually, such a bell is so loud and is so often located near the door that the burglar can hear it quite well. If, after several tries, no one answers, then you've cooperated with him and given him a clear indication that the field is clear.

If possible, relocate your doorbell so its ringing cannot be heard outside the door. If you cannot do this, perhaps you can have it muted somewhat so that it is still inaudible externally. Now, if a burglar pushes your doorbell, he will hear nothing, and will assume of course that your bell is out of order. You might even have a transistor radio playing near the door. In this case the burglar may think you cannot hear the doorbell over the sound of the radio. If your front door has a knocker, have it removed. These make quite a bit of sound and so a burglar can verify if you are home or not. However, as a general rule, burglars usually do not care for knockers because the sound can often carry into an adjacent apartment. For the same reason, they are often reluctant to knock on a door. Both methods make noise, and that is just what the burglar does not want.

MASTER KEYS

In many apartment houses, the management insists on having a master key so they can get into any apartment, if necessary. It is quite possible you may feel less secure knowing that someone, somewhere, has a key that can open your front door. One way out of this dilemma is to use the lock that is opened by a number arrangement, the type that will open only when numbers are pushed in a certain sequence. The usual combination lock has a fixed number sequence, but the doorlock type (depending on the manufacturer) permits you to select your own combination and to change that combination easily and quickly. Thus, you can have a daytime combination known to your building management. During the day, while you are away at work, they can open your door by using the number series you supplied. At night, though, you can change the combination, and so anyone trying to get in then would not be able to do so. This does mean, however, your apartment is still vulnerable while you are away. Before you make any lock changes, discuss the matter with apartment management. It is also better to do so *before* you sign a lease, not after. Once you sign a lease, you are a captive customer and are certainly not in a good bargaining position. Before you sign the lease, you are a customer; after you sign, you are a tenant. And there's a world of difference between the two.

WINDOWS

The fact that an apartment is well above ground level does not mean that windows cannot be used by burglars as entrance areas. Apartment users are sometimes lulled into a false sense of confidence, reasoning that such windows cannot be reached. The burglar is aware that you think this way and that such windows will probably be unlocked. Knowing that they offer easy entrance areas, he will make every effort to reach them.

Bathroom Window

The fact that a bathroom window is narrow, is several or more floors above ground level, does not mean it is secure, particularly if that window is located near a fire escape. Use any of the locking methods described in the preceding chapter. The burglar knows, as well as you do, that bathroom windows are often kept open to ventilate that room and to help keep that room dry. Burglars also are aware that security measures are not extended to such a window and so will make every possible effort to reach it. But because the window is above ground, even elementary security precautions will keep the window locked.

Windows and Fire Escapes

In some states it is against the law to block the exit from a window facing a fire escape. Generally there are restrictions on keeping objects such as plants on the fire escape or outside window sill. The purpose here is for your protection, to let you escape from a fire as easily and quickly as possible. But this lack of clutter also makes it much easier for the burglar. If the fire escape is toward the back of the apartment house, the burglar is less likely to be seen simply because there is less illumination and also fewer people. You cannot put an ordinary metal gate across your window since this would defeat the purpose of the fire escape. But you can now get a gate that has an easily released latch that permits the gate to swing out of the way.

HOUSES VERSUS APARTMENTS

Some burglars prefer apartments to houses. They know how careless people are about front-door locks, so in many cases all they need do is to walk down an apartment-house corridor and try each front door until they find one that is open. A quick ring on the bell soon tells them if anyone is home. If someone is home, they have a convincing story ready. If not, they move quickly into the apartment.

A burglar not only wants to get in, but wants to be sure of being able to get out. And so if he comes in a back window, he may open another window and/or front door to give him a chance for a quick exit. If he comes in through the front door, it is quite likely that even before he starts collecting your possessions, he will open the window that faces the fire escape.

Another reason a burglar may prefer an apartment is that there is a smaller likelihood of occupancy. A house is usually for a fairly large family and is seldom occupied by just one person. On the other hand, there are quite a few apartments that are used by just one person. The average burglar prefers no confrontation at all, but if there is to be a face-to-face meeting, wants the odds in his favor as much as possible. The fewer the people, the better the odds.

APARTMENT LIGHTING

Even though you have an apartment, a burglar walking down your corridor looking for a target of opportunity can glance at the bottom of the door and learn if your apartment is lighted or not. If it is about eight o'clock in the evening and there is no light shining under your door, the burglar may immediately consider your apartment as a likely prospect.

For an apartment, as well as for a home, it is helpful to have one or more electric timers that can turn lights on—and off— at the hours you select. If you go out of your apartment and it is dark, or soon will be, and expect to return within an hour or so, do not leave a dark apartment. Don't turn on your foyer light—no one keeps a foyer light burning for any length of time—but do turn on your living room light. If there is a door between your foyer and your living room, keep the door open so that the light will have a chance to reach the bottom of your front door. If you expect to be away most of the night or all night, use a timer so that the light will turn off at about one o'clock in the morning. A light that remains on all night is as much a signal to a burglar as no light at all.

RADIO RECEIVER AS A SECURITY DEVICE

In addition to lights, turn on a radio when you leave. Keep it playing at low volume, adjusting it so that it can be heard by someone at the door, but not loud enough that it will disturb your neighbors. One way of doing this is to put the radio somewhere near the door. This will allow you to keep the volume low and, yet, let it be heard at the door front by anyone listening there. You can also connect the radio to a timer so that the radio will go on and off intermittently.

ALARM SYSTEMS

Unlike the owner of a house, an apartment resident is carefully restricted by the terms of a lease as to what he may or may not do in the way of making changes in the apartment. The lease usually does not give you the right to install an external alarm system, either on the outside of the building or in your hallway. As indicated earlier in Chapter 4, not all alarm systems are alike. Those that require some wiring changes probably require your building owners' permission.

Buying an Alarm System

When you buy an alarm system, you will probably get a self-stick label you can affix to the door, stating that the premises are electronically controlled. It might seem strange to use a label such as this on an apartment door, but keep in mind that with every burglar you must try to do three things: (1) You must try to discourage him from trying to enter. The warning label does just that. And so does the sight of a well-protected lock, or the sight of a pair of locks, and a peephole. (2) You must make it as difficult as possible for the burglar to get past the barriers you have erected. (3) Once the burglar is inside you must make it as uncomfortable and as potentially unprofitable as you can. You can do this by using a system that turns on an alarm and lights, and by not keeping valuables at home or by having them well out of sight and difficult to locate.

If you are not permitted by your lease to install an alarm system or if you don't want the expense at this time, you can still buy the warning label and fasten it to your front door. The burglar may suspect you do not have such a system, but he will not be sure. And that's the point. If it keeps him from entering, it will have achieved its purpose.

Naturally, you can defeat the label by talking or bragging about it. Don't make it the subject of casual conversation and don't confide in anyone. Just put it on your door, forget about it, and let it do its work. And don't let yourself be questioned or cross-examined about it. Your personal security and safety are precisely that—personal.

One of the greatest problems about security is that you must be conscious of it just about all the time. If you have an alarm system in your apartment, you must remember to set it every time you are home and you must test it once in awhile to see if it is working. This can be quite a nuisance, and so after the novelty of having the unit has worn off, you may be inclined to let it go and to forget about it. An alarm system can protect you only if you give it a chance to do so. The same sort of thinking applies to window locks or to any other locks that must be keyed from the inside. The best thing to do is to form the security habit. Set up a routine for yourself so that you check security at night before you retire and each time you leave your apartment. It is the mental effort in remembering the steps to security that is fatiguing, not the steps themselves. And so, before going to bed at night, check the door and windows. Is the door locked, bolted, and chained? If you use an inside lock with a cylinder, have you closed it with your key? Are your windows key-locked? If you have an alarm system, is it set to the "on" position?

Making Your Own Alarm System

If you do not want to buy an alarm system you can make your own fairly inexpensively, as shown in Fig. 5-6. Know as a pull-trap sensor, it consists of a clothespin made either of plastic or wood, a battery, and a horn.

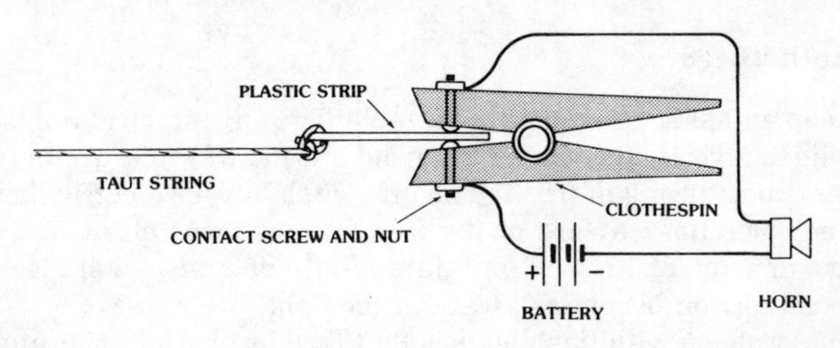

Fig. 5-6.

Homemade pull-trap sensor.

Drill holes in the ends of the clothespin and put a 6/32 machine screw (or larger screw) through each hole, fastening the screws with hex nuts. Wrap a wire around each screw, holding the wire in place with the hex nut. Connect the wires to a battery and a horn, as shown in the diagram. Connections to the battery are not critical and you can transpose the leads.

Insert a strip of plastic between the two screw heads, opening the clothespin slightly to permit this. Connect a string or a length of wire to one end of this plastic strip and then run the string across the area to be protected. Thus, the string, which should be taut, could be strung across your front-door entrance. Opening the door will pull the string and this in turn will remove the plastic strip from between the two screws in the clothespin. Once this happens, there will be a conducting path for current to pass from the battery to the horn and the horn will sound.

There are a few problems with this very elementary alarm system. The first is that once the plastic strip is moved, the alarm will not only sound, but will continue doing so until the clothespin is opened or the battery has run down. Your neighbors may not be appreciative.

An alternative arrangement would be to use one or more lights in place of the horn. In any event, you would need to be very careful not to trip the string every time you enter or leave your apartment.

You can also use this setup to protect windows. Remember, though, it is important to make the pull string as taut as possible since slackness in the string will defeat its purpose.

FLASHLIGHTS

Keep a flashlight on your night table. The advantage of a flashlight is that you will not need to get out of bed and grope around in a dark room trying to find the on/off switch to your electric light, at the same time, aware of the fact there is noise elsewhere in your apartment. Some burglars (but not all) will leave immediately on becoming aware of the light.

The problem with flashlights is that they are battery operated. Batteries wear out whether you use them or not so, about once a month, run a check on the flashlight to make sure it is still operative. If the flashlight uses two or more cells, replace them all. Select a flashlight that is easy to use and is equipped with a switch that lets the light remain turned on without the need for your holding your fingers on the switch. A well-made flashlight is a good investment.

TELEPHONE

You should have a telephone on your night table. Keep emergency police and fire phone numbers pasted right on the phone. Use a touch tone phone in preference to a dialing type. If you do not get many phone calls, check your phone every day or so just to make sure you hear a dial tone, an indication the phone is working.

LAMPS

You should also have a lamp on your night table. If for some reason you cannot, you should have an extension cord with a switch at its end so you can turn on a light elsewhere in the room. It may seem unnecessary to have a flashlight if you have a bed lamp, but the flashlight is also useful if the power should fail.

SECURITY LIGHT

For use in an emergency or in the event of a power failure, an automatic security light like the one shown in Fig. 5-7 is also a good investment. This portable light simply plugs into any convenient wall outlet and continually charges its battery pack. If the power should fail, the security light comes on automatically, drawing power from the batteries. It also functions as a night light or it can be removed from the power outlet and used as a flashlight.

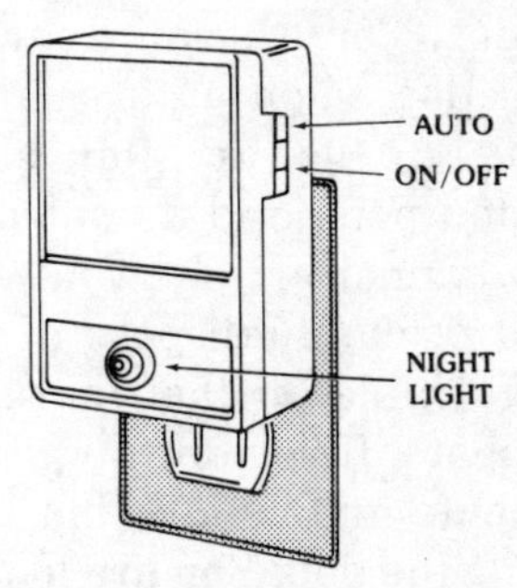

Automatic security light. (Courtesy General Electric)

Fig. 5-7.

DOGS IN APARTMENTS

Keeping a dog (or more than one dog) in an apartment for security is an excellent idea, if you are permitted to do so by your lease, and if you are a dog lover, or at least do not have an antipathy to them. You must also have someone who can take care of your dog if you need to be away all day, and be willing to pay to have a veterinarian check out your dog once in a while. You must be prepared to accept the trouble and bother associated with taking care of an animal. Now those are a lot of demands on you, but there is a payoff in the form of security.

In an apartment, a dog is a living alarm system and in some respects is better, more secure protection than a dog in a private house. In a private house, a burglar can have a direct confrontation with the dog and can soon determine whether the dog poses a threat. Some burglars have a way with dogs and can silence them quickly with food, soothing words, or some heavy instrument, if all else fails. But in an apartment, the burglar cannot see the dog, does not know whether the dog is a miniature poodle or a German shepherd, and has no way of calming the animal. Once the dog starts barking, that's it. Usually, it isn't the dog the burglar is afraid of, it's the noise.

Certain dogs are specifically trained as "attack" dogs for use in commercial establishments. These dogs are vicious, deliberately made so through intensive training. Most people do not know how to handle such dogs, do not know how to exercise them, and certainly do not know how to control them. They do not belong in an apartment.

Oddly enough, the kind of dog you should get depends on the size of your apartment, and the larger the apartment, the bigger the dog. For a small apartment, consider the advisability of a small dog, such as a small fox terrier or poodle. They can be trained to use a litter box, are economical to feed, become devoted pets, and do not require an undue amount of exercise. They cannot provide you with physical protection and a burglar could probably dispose of them with one well-aimed kick. They are valuable for their bark, for their tendency to yelp and yip at strangers. If you decide to get a dog, don't get a puppy. They may look cute and cuddly, but keep in mind your primary purpose in getting the dog. You are looking for security, not love or affection. Unless you are interested in pedigree, you should concentrate on getting a dog that is trained and is housebroken. There is no reason why your dog should consider your entire apartment as his private bathroom.

A dog that is trained can become untrained. Many people spoil dogs just as some parents spoil children. And so it is sometimes necessary to train the apartment resident as well as his (or her) dog. From time to time it may be necessary for you to take your dog back to the supplier for a refresher course.

The dog should not soil the house, and should not regard all objects within it as his personal playthings. He should be gentle with children and must be able to tolerate the unthinking pullings and pushings of toddlers. If the children are old enough, they must be taught that the dog is a living creature, capable of being hurt and abused.

When you decide to buy a dog, buy from a store or dog supplier with the understanding that the dog is trained. Explain your circumstances, the size of your apartment, the number and age of the occupants, and the approximate amount of time the animal might be left alone. Get all the information you can about the dog *before* you buy: how much food to give the dog, what kind of food, how often to exercise the dog, how to maintain the dog's training, how not to overindulge the dog (a common failing), and how to keep the dog in a peak healthy condition.

YOUR INSURANCE

If you live in a high-crime area, you may not be able to get insurance for jewelry or furs. A high-crime area is not only one in which there is a great proportion of crimes, but usually includes quite a bit of surrounding territory as well. In some cases you may be able to get only a limited amount of insurance, or the premiums may be extremely high.

Just because you have theft insurance does not mean you will get full payment for the articles stolen. Most possessions depreciate and all you will get will be a percentage of the original cost, assuming you can prove forcible entry. And that's the catch. If, feeling secure because you have insurance, you barge off and leave your apartment door unlocked, or inadequately locked, and a burglar can get in without making so much as a mark on your door, you may have trouble proving theft. An insurance company will demand proof, and there's no better proof than a door that has been forced. The burglar, though, has enough problems of his own without worrying about you, too. He wants to get into your apartment the easiest possible way, without risk, and without work. He isn't going to use a jimmy if he can use a key. He isn't going to use a jimmy if a plastic credit card will do the job.

Your should also check with your insurance company to learn if it gives premium discounts to those who have installed security devices or who have made various other efforts to make their apartments secure against theft. Your insurance company may also be able to advise you of which security steps it recommends.

Chapter 6

How To Make Your Car More Secure

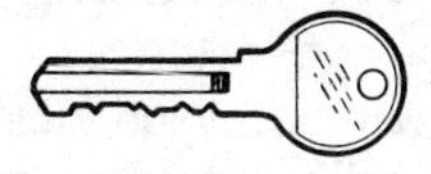

The number of car thefts in the U.S. is staggering, with more than a million cars stolen each year, or about one car every 32 seconds. There are not that many criminal professionals in this business, since approximately 80% of all car thefts are done by amateurs, yet they represent the main force in a business that can easily call itself a billion-dollar industry. It is estimated that a car thief can enter your car in about 10 to 15 seconds, and that includes driving-away time. As far as most cars are concerned, they are either not protected against the criminal or pitifully so. If a truly professional thief wants to steal your car he can do so.

Next to our homes—and often on a par with them—our most prized possessions are automobiles. The remark that "Americans have a love affair with their cars" would be more humorous were it not such an accurate statement.

Our romance with the car is due to any number of reasons. Some of us have become exurbanites, and this removal from the city immediately means dependence on the car for shopping, for transportation to the job, and for social activities. We have been so urged by car manufacturers to buy, that not uncommonly many have become two-car families. And the possession of "wheels" is now the great dream of young people.

CAR THEFT VERSUS HOME THEFT

But the car is not only a great convenience and a pleasure, it is a prime target for all those looking for the easy buck. Just consider the theft opportunity a car presents and compare it with the same opportunities for stealing from a house or apartment. With the house or apartment the thief often does not know whether someone is home or not, and this is a risk he must assume. With the car, there is no such problem, for the thief need merely look in the window. And when he breaks into a home or apartment, the burglar must take a chance on whether or not there is something worth stealing. Yes, there are some burglars who are satisfied with a haul that will net them a few miserable dollars. But for many burglars, such a return isn't commensurate with the risk. With a car, however, any expert car thief can evaluate fairly well what he can expect to get. He can recognize the year and model of the car and a quick look will soon tell him something of its condition. If the paint job is good, if the tires have an adequate amount of tread, if the interior doesn't appear too worn, then the burglar knows almost as much about the car as the owner.

There is one more feature about the car that makes it so attractive a target for theft. Most car-door locks are pathetic. An experienced thief can punch through the car lock almost as fast as its owner can open it. In some cases the thief may not want to do this since he doesn't want to damage "his" property, and so with the help of a coat hanger, is able to release the door, usually in rather fast time. Cars that are illegally parked are opened this way by traffic police, and quite obviously they do not find a locked door much of a barrier. And in many instances, car owners have the habit of slamming their car doors without checking to see if this action has locked the door or not. Similarly, many lock their doors and leave windows open.

Consider also that while apartment dwellers may pay some attention to their neighbors, car owners do not even have the slightest interest in any adjacent car. All the car owner wants is to have a parking space, not to have the entrance or exit blocked by other cars, and not to have the car scratched or otherwise damaged. And yet the same person who cleans the car with a lint-free chamois every Sunday casually and unthinkingly exposes that pride and joy to thieves and vandals.

TYPES OF CAR THIEVES

There are probably as many varieties of car thieves as there are burglars, but they can be grouped into a few main categories:

(1) the professional, a thief who steals to order; (2) the opportunity thief who will steal any car, if it is convenient for him to do so; (3) car parts thieves, also known as car clouts or boosters; (4) the joy-rider who steals for the satisfaction of having a set of wheels, even if for just a short time; and (5) the vandal, who doesn't actually steal a car but is simply intent on damaging it. In some cases, thieves may also commit acts of vandalism if they are unable to remove the equipment they want.

The Professional

If there is such a thing as an elite class in car theft, the steal-to-order thief is it. By prearrangement with a larcenous-minded customer (or the thief may be working for a group) the thief is told to be on the lookout for a model and year of car. He or she may even be informed of a preferred color. Once such a car is located, the thief will stalk it to learn something of the driving habits of its owner. With this information on hand, and since the average car is so easy to enter and drive away, the car thief doesn't find it very difficult to make delivery.

The thief who gets away with a car on order does have the problem of supplying a new set of plates and registration. He or she can get these from car wreckers. Since we have an individual state rather than a Federal registration system, it requires no great effort to transport the car to some other state. Sometimes the car is shipped to another country. There is a brisk overseas trade in stolen cars.

The Opportunity Thief

Next to homes, automobiles represent the largest single investment made by most American citizens. Yet, most people are notoriously careless with their second most valuable property.

Now What About Opportunity?

Out of all the cars that are stolen, about three-fourths of them were left unlocked by their owners, and almost half even had the keys left in the ignition locks. The opportunity thief does not even need tools. All one has to do is to walk through a parking lot, conveniently provided by shopping centers, or along any typical residential street, look inside the parked cars, and quickly try the doors.

Most stolen cars are eventually returned to their owners, but statistics of their return do not reveal the whole story. In many instances valuables inside the car disappear, never to be found, or the trunk is forced open to yield its contents. Some people even put valuables in the glove compartment. Any self-respecting thief can dispose of the lock on that compartment very quickly, assuming the lock is used by the owner; most often it is not. The 16 percent of the cars that are never returned represent an annual loss of over 140 million dollars. Even when the vehicles are recovered, they may very well have been stripped.

The Car Parts Thief

Don't take it personally, but some thieves think more about your various car parts than they do about the entire car. There are basically two kinds of car-parts thieves. One is the thief who has a car model identical to yours and needs parts for his car, parts he is either unable or unwilling to buy. He may not only strip your car of various accessories, but may take others as well to keep an inventory on hand just in case he should ever need them.

The second type of car-parts thief makes a specialty of stealing car parts. In some instances he can earn as much for these parts as he could for the entire car, but with much less risk. Automobile tires, grilles, front fenders, hoods, bumpers, and batteries do not carry identification numbers and so for these items the thief can find a ready market.

The Joyrider

The joyrider is most often a young male, although some young females are getting into this scene. The joyrider may have his own car, but may take yours on impulse, just to "try it out," or he may be on a date and wants to make an impression, or he may have neither a car nor a license but is absolutely convinced he is tops as a driver. Whatever the reason may be, your car will disappear for a number of hours and during that time it will be in the possession of someone with little driving experience, willing to take risks, prone to driving at high speeds, and inclined to operate the car using driving techniques that will do the car no good. Jack-rabbit starting (burning rubber) is just one example.

Oddly, most car thieves are not professionals, but amateurs. And about two-thirds of these amateurs are teenagers. What you are up against is a young kid who is out to take whatever he can and who has absolutely no thought for the consequences. And since he cannot or will not think ahead, he will take the first car that is readily available. For the teenager an available car is one whose door is left open with a key in the ignition. There is no real reason for him to break into a car when there are so many available without effort.

Note this combination: unlocked door and key in the ignition. But if you lock the door and leave the key in the ignition, then you are in even greater trouble, for now you have put irresistible temptation in the path of the teenager thief. His reaction is predictable. He'll smash a window with whatever he has available—a brick, an empty bottle, a hammer handle, anything. And then he is in and away in a moment.

The problem with the joyrider thief is not that you will not get your car back. You probably will. But it may no longer be recognizable. The joyrider thief is highly accident prone, firmly believes he is invulnerable as long as he is behind the wheel of a car, and may very well be aware that he or she is a juvenile and so will be treated leniently if caught. He or she may also realize that the average car owner may be so delighted to get the car back that he may not be willing to press charges.

The Vandal

Next to the teenager joyrider, the vandal may very well be the worst of the car thieves. He is really not a thief; generally he is someone ranging in age from the teens to the mid-twenties who enjoys destruction. He will bend your whip antenna out of shape or may bend it back and forth until it breaks and then walk off with it. Or, he may take a stick with a nail protruding from its end, and then walk along a line of cars scratching the paint. Or, he may slash tires. Or, smash windows. He doesn't steal, but he can do as much damage as a thief. The problem with the vandal is that his behavior is so senseless it is unpredictable. And while a car can be protected, at least to some extent, against a thief, there isn't too much you can do about the vandal. We now have antennas that can be recessed to prevent someone breaking them off, but there are no slash-proof tires or scratch-proof paints.

NEW CAR VERSUS OLD CAR

Don't get the idea that car thieves are interested only in brand-new cars. Quite often that is a matter for the professional. The joyrider type of thief does not want a new car for it attracts attention. He prefers an older car, knowing it will merge much more easily into the passing car parade. Boosters will take anything they can get their hands on, parts such as batteries, wheel covers, radiator grilles, CB radios, stereo receivers.

VIN

Every car has a vehicle identification number or VIN. This number is located in one place in your car where it is easily seen and in another, hidden location. Vehicle ID's can be changed. The car thief does have the problem of locating the hidden number. Car thieves, working as a group, can purchase a late model car of the kind they intend to specialize in, and then strip it, bolt by bolt, until they locate the hidden vehicle identification number. From that point on, they can steal similar model cars with the assurance that they can modify all identification numbers. Forging registration papers, insurance papers, or any other documents required by the individual states, is comparatively easy. With such a setup in active operation, the car ring is equipped for wholesale theft.

Time Payments

One of the unhappier aspects about having your car stolen is that the theft does not release you from the need to continue to make payments if you bought your car on a time-payment plan. You are still responsible for making those payments. The finance company or bank that loaned you the purchase money for the car originally is not responsible for the theft of the car, and does not share with you in its loss. And so now you are in the unenviable position of making payments on a car now in someone else's possession. If your car is stolen, you do not win. You lose, to a greater or lesser degree.

The fact that your car was stolen does not absolve you of responsibility for it. The point is, you are still the lawful owner of the car and if the car is involved in an accident, even if driven by a thief, you can still be sued, and depending on various laws in the individual states, may be forced to pay. And so, even if your car is an old one and its theft is of no great immediate concern, you may still be forced to pay far more than the car is worth.

Contributory Negligence

Your "guilt" in an accident caused by the person who stole

your car may be predicated on your contributory negligence. If the thief is driving the car with your car keys and your registration, it would be easy to prove that you practically invited the theft of the car. But if your car is equipped with various devices to hinder or balk car theft, and if you can demonstrate that you did everything humanly possible to prevent the theft, then you may be in a somewhat better legal position.

YOUR CAR REGISTRATION

One of the ways in which the car owner cooperates with the thief is to leave the registration of the car in the glove compartment. And some owners have the unhappy habit of keeping a duplicate set of house keys in the same place. So now the car owner may not only lose his car, but has literally given the thief an invitation to rob his home as well. If, for example, the car is in a no-pay parking lot, the thief will realize he has just about enough time to visit the victim's home. Even if the keys aren't in the glove compartment, the registration does have the name and address, and quite often that is all that is required.

INSURANCE

Insurance is part of the reason why car owners cooperate so wholeheartedly with car thieves. What we should realize is that the theft of someone else's car affects all of us. Insurance rates are predicated directly on the number and value of total thefts. If these are high, premiums for theft insurance are correspondingly high, and so even if your car is never stolen, you are still paying for someone else's negligence.

Insurance does *not* mean the insurance company is going to give you enough money to pay for a brand-new model of the car you have been driving. As a car gets older, its value decreases, with the drop in value the greatest amount for the car's first year of life. As a matter of fact, there is a serious drop in value the moment you drive a car out of the showroom. This value decrease does not mean the car is worth less to you, simply that you would get that much less for it in the resale market. Now add to this the fact that theft insurance also includes a certain amount of deductible (the amount the insurance company takes right off the top when evaluating the market worth of your stolen car). And so, what you do get from the insurance company is much less that the value you have mentally put on the car, and may often amount to just enough for you to make a down payment on another car.

Some insurance companies may decrease the amount of your premiums, depending on how wholeheartedly you cooperate by equipping your car with antitheft devices. In some instances, an antitheft program could decrease your comprehensive insurance costs up to 15%, but for a precise amount you will need to check with your insurance agent. But, even if you cannot get a lowering of insurance rates, equipping your car with antitheft devices will increase the resale value.

No insurance company likes to pay out to losers, people who are regularly the victims of thieves. If you tend to be burglary prone you may find it more and more difficult to get insurance coverage or else you may suddenly realize your premiums have skyrocketed.

Many new auto-insurance policies specifically exclude car stereo and CB units unless there is an extra-fee rider attached to the policy or unless the equipment is installed at the factory and thus is considered an integral part of the car.

There is confusion among claims agents on whether an in-dash installation after original sale of the auto would qualify as "factory" installation under the insurance policies. The reason the insurance companies are so tough about insuring stereo and CB components in cars is the high incidence of theft. If your present stereo and/or CB is covered by original equipment insurance, better check with your insurance agent if you plan to install new components. Your policy may not cover the new units.

HOW TO PROTECT YOUR CAR

Before you try to come to a decision that you do need car protection, and just about all cars do, consider whether you want to protect the entire car, or just some car components, such as your grille, or wheel covers, or whether you want to do both. But even before you buy antitheft equipment there are antitheft measures you can take.

There are a number of ways of protecting your car and a variety of commercial devices available, but just as a house or apartment can be invaded by a very determined thief, so too is your car vulnerable to someone who has made up his mind to steal it. However, there are some important factors in your favor. One of these is that the thief is just as disinclined to work as anyone else and so will pass up a car that obviously has some protection in favor of one that looks like "easy pickings." Another factor is that the car thief, like the apartment or house burglar, likes to work in privacy and preferably in the dark. The need for privacy and darkness, though, is gradually giving way as thieves are beginning to realize that most people aren't interested in activities involving someone else's car—just so long as it isn't their own.

Here is a list of things you can do to protect your car:

1. Do not leave your car key in the ignition. And don't take the key out of the ignition and put it in the glove compartment. Take the key with you and make sure it does not go out of your possession. Do not keep your car keys—and that means ignition and trunk keys—on the same key chain as your regular house and office keys. It is much too easy to hand a parking attendant your entire key chain. When parking in a commercial parking lot, leave your ignition key only. Remove your trunk key from your automobile key ring before driving into a commercial parking lot.
2. Make sure you have two sets of car keys, ignition and trunk, and let your wife hold the second set. In this way, if you do leave your car keys in the ignition but lock the door you can arrange to have the door opened quickly to retrieve your keys. Keys in the ignition present a most tempting target.
3. Don't leave your registration in the car, and if the laws of your state also require other papers, such as an identification card or proof of insurance, keep them in your wallet. The only papers you should have in your glove compartment are those the laws of your state require.
4. Don't assume your driveway is a safe parking area just because it belongs to you (Fig. 6-1). Your car can be stolen from your driveway just as easily as from the street. Put your car in your garage. And don't leave the garage door open. Do this even if you are planning to use the car again at some later hour. When your car is in the garage, behave as though your car were out in the street. This means closing all car windows, making sure the ignition key is in your pocket, and that all doors to the car are locked. A car door that is shut isn't necessarily locked. If your garage has a window, keep it closed with a key-type window lock.

Fig. 6-1.

A car parked in driveway or carport is not automatically protected.

5. If you have more than one car and you have just a single-car garage, park the less desirable car directly in front of the garage door, and not out in the street.
6. The only things of value you should keep in your car trunk are your spare tire and the tools you need for changing a tire. A car trunk is easy to open. However, if you are out on a trip, hide everything you can in the car trunk. This doesn't mean it is safe in the trunk, but at least the thief will not know whether it is worth his while to make an effort to open the trunk. If he sees merchandise all over the car seats, just waiting to be stolen, he may accept your kind invitation. If you have merchandise in your car, covering the goods with a newspaper or old blanket is futile. You can be sure the burglar is aware that you are as lazy as he is.
7. If you have an electric pencil, mark your name somewhere on a hidden metal surface of your car. This can be inside the door, on the rear of a bumper, or inside the hood. If you don't have an electric pencil, use a sharp-edged tool such as an awl or pick or even a screwdriver. Use an electric pencil to engrave your name and/or social security number on the side or back of electronic equipment that you have had installed in your car, including CB gear, stereo sound equipment, or a radar detector. Make a record of their serial numbers, model numbers, and the names of their manufacturers. Do not keep this information in the car.

Instead, attach the list to all the sales slips and put it together with your other important papers.

8. If you are planning to buy a new car, become security-conscious. Give preference to a car that has a buzzer to remind you that the car key is still in the ignition.
9. If you are planning to buy a used car, keep in mind that if you buy the car from an individual you must return the car to its original owner if it is a stolen car. The responsibility is yours. Be careful about ads placed by individuals offering to sell a car at what seems to be a tremendous bargain. This is one way in which thieves dispose of cars. However, if you buy a used car from a dealer, then it is the dealer's responsibility to make sure he doesn't sell you a stolen car. If he does, and it is repossessed, you are entitled to a refund.
10. Try not to park your car in known high-crime areas.
11. When parking, try to park your car in the most lighted area.
12. Operators of open-air parking lots are generally not responsible for the security of your car. The extent of any possible responsibility varies from state to state, but you can safely assume it doesn't exist. Nor are open-air parking-lot operators responsible for the contents of your car, and that includes the trunk.
13. It is better to park in an open-air parking lot than out on the street; it is better to use an indoor parking building than an open parking lot.
14. A dog in a car is good security but may be tough on the dog. If the day is hot and the car is parked where the sun can reach it, this would be an act of inhuman cruelty. Use a dog when you plan to be away a short time.
15. If your car becomes disabled while you are traveling, stay with the car. Hang a large white handkerchief from the antenna of your car. This is now generally being recognized as a car-distress signal. Try to get your car towed to a garage as soon as possible. You may regard towing rates as very high, but these rates are much cheaper than having your car completely stripped. Strippers can cannibalize your car almost as rapidly as a thief can steal it. An obviously abandoned car is a direct invitation to theft.

16. When parking your car, whether in a parking lot, garage, or out in the street, never let it be known how long you will be gone. Some parking attendants do ask, so make your answer: "Not long." If you must park in an open parking lot, try to select one that has just a single entrance and exit. If you use the lot regularly, get to know the attendant's name and use it very time you park. Mention your own name. Sooner or later, despite the large number of cars parked, he will associate you, your name, and your car.
17. Do not leave your car parked overnight in an airport parking lot. These are often open-air types, are rarely fence enclosed, have several entrances and exits, and are so large that no attendant could possibly remember you or your car. A thief can drive in a piece of junk that can barely move, pick up a parking ticket, and then use that ticket for driving out with your car.
18. The best place for your car at night is in your own locked garage.
19. If you commute to and from a bus or train, always park your car in company with other cars and not by itself off in some odd corner of the parking lot. Yes, your car is more likely to get scratched or otherwise damaged by other parkers, but it is also less likely to be stolen.
20. If, as you approach a red light, you suspect your car is being followed or if a group of pedestrians approach your car, one or more from each side and possibly from the rear, and you are in what appears to be a somewhat deserted area, go through the light. Cross the intersection, make a left or right turn. Do anything, but get away from the neighborhood, and as fast as you can. If you need directions, do not stop. Keep going until you come to a better district.
21. If you know you are going to drive through a high crime rate area, or, if you are in a strange city and are driving through what appears to be a rundown section of the city, keep your windows closed and your doors locked. Get out of the area.
22. If you must take a taxi to a high crime area, ask the cab driver to wait until you have entered the building you are visiting. Before you leave, telephone for a cab and wait until the cab arrives before leaving the building.
23. When you park your car at night, don't turn off the car's

lights until you are fully ready to leave the car. If it is night and you have packages in your trunk, other than perishables, let them stay there until daytime. Try not to park on a deserted street, even if you have been looking for a parking space. Always try to park as close to your home as possible.

ANTITHEFT ACCESSORIES

An antitheft accessory is any device which will hinder the theft of a part of your car or the car itself. There are quite a number of such accessories you can use for your car, none of which will actually prevent a car from being stolen. However, the more difficult you make the job of theft, the less likely it is that your car will be stolen. There is no reason for a thief to spend two hours trying to take your car when the car immediately adjacent to yours can be had in a matter of minutes.

Interior Knobs

Older cars are equipped with a flanged door knob on the inside. The unit is somewhat like a button and is designed to move up and down. In the down position, the door is locked; open when the knob is up. This type of knob makes it easy for you to unlock the door, but it is almost as easy for a thief. All one must do is to work a length of stiff metal wire—a coat hanger does the job very well—past the rubber gasket of the window. With the end of the wire fashioned in the form of a hook that can engage the flanged doorknob, the thief can open your door quite easily. Once inside your car, no one will pay much attention. The only pressure the thief will experience is your possible unexpected return.

Not only thieves, but policemen as well, use this technique for removing cars that are illegally parked. Once your car is towed away, the penalty is quite often a rather stiff fine.

To minimize this action, you can replace your present interior flanged-type door handles with round knobs. This does make it a bit more difficult for you to open your car doors, but it also is a theft deterrent.

You can also get a loose fitting sleeve that can be positioned right over the knob. When the wire hook (generally fashioned from a coat hanger) slips over the knob, the sleeve, looking something like a cap, slides up and down, keeping the hook from getting a grip on the knob. You can also replace the knobs with unflanged types.

Modern cars have eliminated the knob-type lock, replacing it with a knob that is small, smooth, and straight. As a result, the wire hook cannot get a grip on the knob. In some cars, the knob is completely eliminated and is replaced by a locking-type door handle.

Wheel Locks

Tire thieves don't bother just stealing tires—they take the entire wheel. And so what you lose is the hub cap, tire, and wheel. Multiply this by four and the replacement cost can easily be several hundred dollars, usually more. In some instances the thieves, who often work as a team, aren't all that gentle in lowering the car to the ground, and so you may have some damage to pay for as well.

But that isn't all. Car strippers, seeing the car minus its wheels, may very well assume that the car has been abandoned, and so will be over and through your car like a horde of locusts, removing the battery, radio, generator, windshield wipers, and anything else having some possible use or resale value.

But that isn't all. Your car is now stripped and apparently abandoned. But its windows are still intact and to some teenagers this is an intolerable situation. They remedy this easily with just a few well-aimed rocks. In the short time you were away, your car could have been reduced to a useless pile of junk. To add insult to injury, you may have to pay to have it towed away, depending on the municipal laws and ordinances of your town or city.

To avoid starting on this chain of events, you can equip each of your wheels with a wheel lock. The wheel lock replaces one nut on each wheel. Get a set of four so you can have one for each wheel. However, even with wheel locks you will still be faced with a problem. You must always be sure to have the wheel-lock key with you when you drive. If you get a flat tire and you do not have the wheel-lock key with you, then you must begin removing the lock forcibly—and that in itself can be quite a job. Don't be tempted into leaving the wheel-lock key in your glove compartment or behind the sun visor. Those are exactly the places the thief will look for it. Keep the key on your key ring or hide it in the car and also give your wife or some other member of the family a duplicate key, or keep a duplicate key at home in a location known to other members of your family. In that case, a telephone call may help solve your flat-tire problem.

Battery Lock

Some thieves specialize in the theft of batteries, for these can be readily sold in the used-battery market. Stealing a battery is a theft that can be done in just a few minutes, and to those who walk by, it does look as though the driver is simply having some engine trouble. The thief just lifts the hood, disconnects the battery terminals, and then raises the battery up and out.

Without the battery, of course, your car just isn't going anywhere. This means you will probably have to pay for towing plus the cost of a new battery. In some cases, cars that have been stripped this way have also received overtime parking tickets.

To protect the battery you can equip it with a battery lock. Generally, once a thief lifts the hood and sees the battery lock he will move on to another car not so equipped. Why should he work, possibly attract attention, and run any kind of risk when there are so many cars with batteries, readily available, and equally ready to be removed?

Hood Lock

The trouble with the battery lock is that it protects just this one car component. A thief that lifts your car hood may not only walk away with the battery, but the generator as well. With the hood lifted, it is easier for him to jump the ignition and get the car started. A hood lock can minimize this risk, but the trouble with hood locks that are bought as accessories is that they usually look so unattractive.

Some cars come equipped with under dash hood locks that are controlled by the driver. These are less visible than accessory hood locks, but are not as effective. Anyone who can get into your car can easily release the hood lock. The accessory hood lock, consisting of a chain and lock, is not as easily defeated.

Steering Wheel Lock

You can get a steering-wheel lock that fastens the wheel to the brake or to the gear shift. One type of lock consists of a long metal bar running between the wheel almost to the floor board. It can be a nuisance to unlock and to store, but it will protect your car. Before stealing a car, a thief with at least some experience will look inside your car. If a steering-wheel lock is seen, it is probable he or she will be sufficiently discouraged to move on to another car not so equipped. The advantage of a steering-wheel lock is that even if the thief manages to jump the ignition, he is just not going to go anywhere, at least not in your car.

The steering-wheel lock, the hood lock, and the various other component locks all present problems to a greater or lesser degree. You can drive if you lose your wheel-lock key or your battery-lock key, but not if you misplace the key to the steering-wheel lock. It will take hours with a hacksaw to cut through a steering-wheel lock.

Supplementary Ignition Switch

You can install a switch that connects in series with your regular ignition switch. This means both switches must be turned on before the car will start. Since the supplementary security ignition switch is so easy to use—you just push it with your finger—quite obviously its value lies in the skill with which you can hide it from the thief. Unfortunately, there aren't too many places in a car where things can be hidden. You can put the switch under the dashboard or in the glove compartment, but those are two places the thief will be sure to look. Further, an experienced thief can trace the ignition wires quite easily.

Still, the supplementary ignition switch is helpful. If you leave your keys in the ignition, no opportunity-type thief is going to bother with your car once he tries the regular ignition key and sees that it does not start the car. Also, many thieves limit the amount of time they will spend in stealing a car, thinking that the longer they are in the car the greater the chances of detection.

How to Prevent Grille Theft

The grille covering the radiator of your car can be made of metal but is usually plastic. Should you suffer the loss of the grille from your car due to damage or theft you will be unpleasantly surprised to learn how expensive these ornaments are.

Grilles are often kept in place by plastic screws, generally about a half dozen. These screws, not equipped with nuts or washers, can all be removed in less than two minutes. To minimize the chances of theft, replace the plastic screws with machine screws and equip each screw with a nut. After the screws are in place and the nuts have been tightened, use a pair of pliers and bend the free end of each screw. As a result, the thief will be unable to use a screwdriver to remove the screws. He can still use a pair of cutters in an effort to get the scews out, but this takes time. Further, you will be forcing him to use the cutters in an awkward position, assuming he was farsighted enough to bring along such a tool.

Protecting Your Gasoline Supply

One type of thief, not mentioned earlier, is the least demanding of all and is the gasoline thief. All he wants to do is to siphon the gasoline out of your tank into his, or, more usually, into some sort of container.

One way you can protect yourself from gasoline theft is to use a keyed gas cap. These are made for various makes and models of cars and an automobile supply house should be able to sell you one. These locking gas-cap devices require a key to open, but not to close. To lock such gas caps, just press down on them when they are in position and the device will lock automatically.

When you buy such a security device, the key will probably be marked "gas." This will help you identify it at once since you will need to open the lock when you buy gasoline. If the key isn't so marked, it would be helpful if you did so. Keep the key on the chain immediately adjacent to your ignition key. You'll need both to go anywhere.

Protecting Your Auto Radio Antenna

Late model cars have the antenna built into the windshield, thus protecting the antenna against vandalism. Antennas are so inexpensive compared to other auto components that there is no way in which a thief can dispose of them. The exception is the motorized antenna, but these are relatively difficult to remove. However, the wire-whip antenna does have some attraction for vandals. By flexing the antenna back and forth along its base it can be broken and removed. The thief then has a steel whip which he can use for other purposes.

To discourage vandalism, use an antenna that has a spring base. This makes flexing along the base useless. The single-element whip with a spring base is best as far as vandalism is concerned, but a telescoping antenna is superior if you want both am and fm reception. Unfortunately, this type of antenna is also attractive to vandals who can manage to break off one of the elements. One method of protection is always to keep the antenna in its fully closed position, with the elements nested inside each other. Extend the elements before you start driving, retract them when you park.

A car and its equipment are not automatically protected by being parked in your driveway or under a carport as shown in Fig. 6-1. The most secure place for your car would be in a garage with the doors closed and locked from the inside and with the car windows rolled up completely and the car doors locked.

Protecting Communications and Entertainment Equipment

Radio equipment for a car consists of Citizens' Band transceivers and stereo sound components. Your car may also be equipped with a radar detector. All of these components are attractive to thieves and find a ready market.

Components such as radio receivers, equalizers, and main amplifiers are generally held in place by a pair of side brackets. Self-tapping screws are used and while this simplifies installation, it also facilitates removal.

There are several things you can do. One is to add another bracket at the rear of the component. This does not prevent removal, but it does delay it. Another is to spoil the slot of the screw head, making it impossible for the thief to use a screwdriver. Sometimes the screws are combination slot head and hex head. This means they can be inserted or removed with a hex nut driver. This is the easiest type of screw for the thief to remove, so replace them with screws that have one-way slotted heads only.

Still another method is to use a cover plate that is mounted over the face of the receiver. The cover plate is equipped with a special base that replaces the usual trim plate. Fitting snugly over the base is a cast aluminum cover with more than ⅛-inch wall thickness, locked into place with a 7-pin tumbler lock (the same type used to protect vending machines).

While you are in your car, you can keep the cover off to be able to operate the receiver. If the receiver is equipped with a cassette-tape deck having automatic reverse, you can listen with the cover on. However, you will still need to remove the cover when you want to change tapes. Fig. 6-2 is a photograph of the Audio Safe showing the cover of the unit before being put into position. Fig. 6-3 is an exploded view showing how the protective cover is installed.

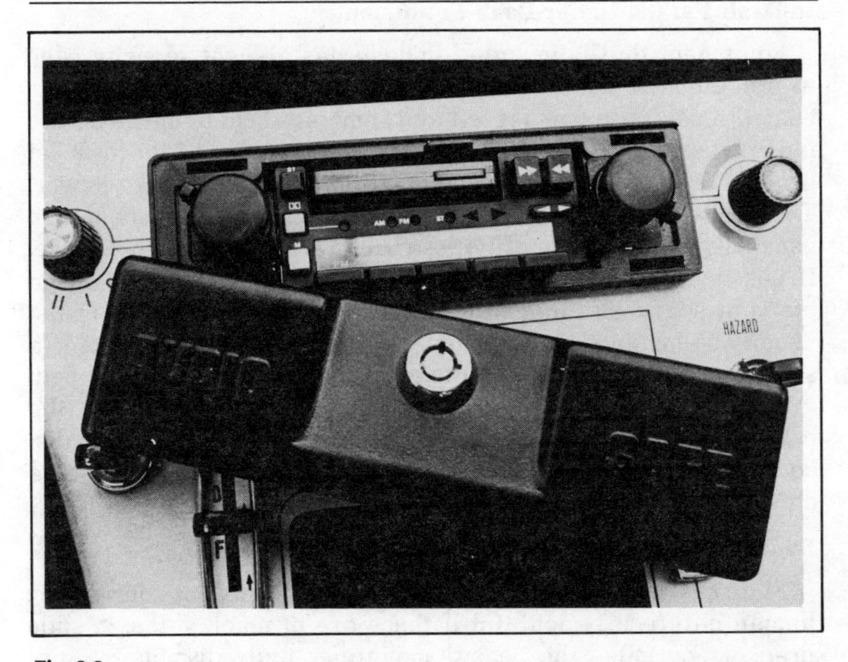

Fig. 6-2.

Audio Safe fits over stereo equipment. (Courtesy Burbank Enterprises, Inc.)

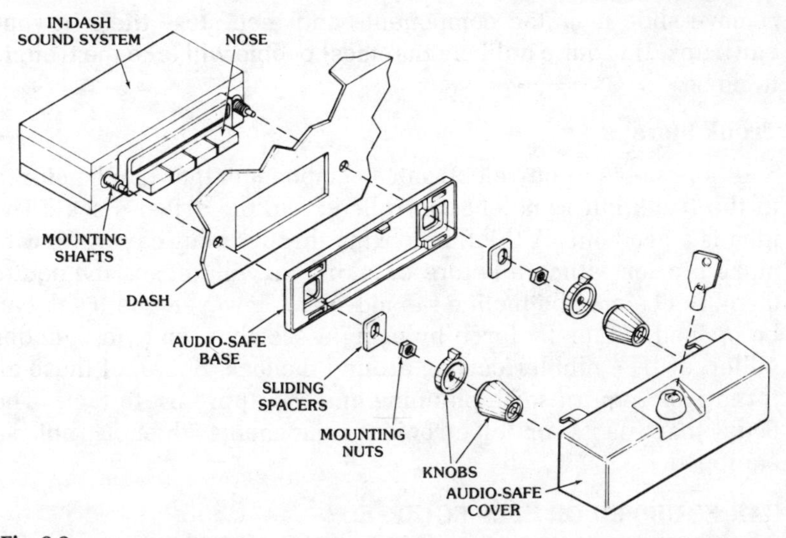

Fig. 6-3.

Audio Safe cover installation. (Courtesy Burbank Enterprises, Inc.)

In-Dash Versus Under-Dash Equipment

Equipment that is mounted in-dash has a slight security edge over equipment mounted under dash. For in-dash components, the controls, usually a pair for left and right, are held to the dash by a pair of hex nuts. These nuts must be removed and the bracket screws undone to get the component out. This does take longer. The knobs, mounted on the control shafts of the component, are generally slide on/slide off types. These are very easy to remove. If possible, replace these knobs with the kind that are held in place by set screws that fit into the knob. These set screws must usually be fastened, or unfastened, by an Allen wrench. This is a small, lightweight tool, but quite often the thief is not so equipped.

Various wires and cables come out of the back of the components. These are often plug in types and so the thief can disconnect them easily enough. If not, he will use a pair of wire clippers.

Slide Brackets

Slide brackets are offered as a security device, but considering human nature it is doubtful if these are of much value. A slide bracket permits the easy mounting and dismounting of components. The theory is that when you park for any time, as during a shopping expedition, you will take the time and trouble to remove slide mounted components and then store them in your car trunk. It is quite unlikely that most people will take the trouble to do so.

Trunk Storage

It is possible to buy electronics components that can be stored in the trunk but remotely controlled from the driver's seat. The idea is a good one. A thief, in looking through your car window to make a prior estimate before breaking in, will not see the equipment, hence may be inclined to move on. However, the trunk can be opened, either by force by using a crowbar, or by a cylinder puller, or by a nibbler cutting around the lock. But all of these alternatives require some planning and the appropriate tools. The usual car parts or electronics components thief is not so equipped.

CAR EQUIPPED OR SELF-EQUIPPED?

When you buy your next car, should you get one that is

security-equipped or is it better to do your own? There are advantages and disadvantages to both methods. The security-equipped car saves you the trouble of getting and installing various security devices; but if these are available as options you will be paying top dollar for them. However, they will save you the time and trouble of doing your own installation or having someone do it for you. If you trade in the car, they may add to the value of the trade-in, but you usually cannot remove these built-in gadgets for your new car.

If you decide to install your own security methods, you have an opportunity to shop for devices that will give you the maximum protection at the lowest purchase and installation cost. Further, you will not have to accept what the car manufacturer supplies, but can select those devices that seem to offer the greatest security. You can also remove them for use in your new car if you decide to sell your old one. Installing your own car-security measures does mean you must shop and spend time in installing, or paying someone to do it for you. Auto-security manufacturers offer a variety of do-it-yourself kits that contain complete instructions, hardware, necessary wire, plus any accessory materials.

AUTO ALARMS

It might seem that one of the best ways to protect your car would be to install an alarm system, somewhat similar in concept to the alarm setup in your apartment or house. But before you do so, you should be aware that in some areas a car alarm must shut off automatically, and is required to do so by law. A thief may set off your alarm and be frightened away by it, but you may get a ticket for violating an antinoise law. Consider these possible case histories:

1. You have parked your car, set your car alarm, and gone to a movie. A thief, attempting a robbery, trips the alarm system and is frightened away by it. Yes, the burglar is foiled, but by the time you return to your car, your battery may be dead, or so weak you are unable to start your car.

2. You have parked your car, set the car alarm, and are off shopping. A thief opens your car hood, trips the alarm, and is frightened away by it. The screaming alarm, though, has attracted some residents in that area who promptly blame you (in absentia) for the continuous noise (and not the thief who tripped

the alarm). In exasperation, they cut wires left and right under your hood in an effort to stop the alarm. Once they accomplish this, they leave, but the hood may still be open. This attracts opportunistic thieves who walk away with your car battery.

To prevent either of these possibilities select an alarm system that will turn itself off after a predetermined time lapse, and will then automatically reset the alarm some minutes later.

Siren Recycler

Many auto alarm systems were sold prior to the establishment of antinoise ordinances. Since, in some areas it is now against the law to have continuous alarm operations, you can modify your present alarm system if it is this type by inserting a siren recycler. This will fit all makes of siren alarms, both the relay and solid-state sensor types. In case of alarm, a recycler will shut the siren off after five minutes and will then automatically re-arm the system, thus offering protection from renewed intrusion.

One other caution: If you have an alarm system, prominently display the telephone number at which you can be reached, so that local police can reach you by phone if the alarm is triggered. In some areas, the law requires that the alarm be equipped with automatic shutoff and that you have your phone number displayed where it can be readily seen.

There are many kinds of alarms and they make all sorts of sounds. You can get a buzzer, a bell, or a siren (either continuous tone or whooping). You can get an alarm that sounds only if the doors are opened, or a type that will protect the doors, the hood, and the trunk. Usually an alarm is triggered or set off by any opening that disturbs a dormant electrical system. Alarms can also be activated when a thief opens the glove compartment, depresses the brake pedal, turns on the car lights, or uses the cigarette lighter. The car owner can use a secret switch or dash keylock to activate the alarm. Generally, there is a time delay so the car owner can get to the switch and disable it. Of course, if the owner forgets to turn off the alarm switch he or she will be reminded about it quite loudly.

Alarm systems can also be used for recreational vehicles, campers, motor and trailer homes, and boats. Some come equipped with a test/panic button to indicate an emergency condition or to test the alarm. Alarms are generally 12-volt types and operate from a standard car battery.

Some auto thieves will try to defeat the alarm system by jacking up the front wheels and towing the car to some location where an alarm can go on and stay on until the burglar manages to defeat it. A car can be rolled up an inclined plane right into a waiting truck, or else hoisted by its front wheels and towed away. To defeat this technique you can use a pendulum switch. When any door, hood, or trunk is opened or when the car is moved, the pendulum switch will close a pair of contacts, sending current to the alarm. Not all alarm systems are equipped with a pendulum switch and so not all offer this additional protection.

Many thieves do not like to work under the conditions imposed by an activated alarm. Yet, some burglars are highly persistent. If they can manage to get into the car fast enough, they may be able to find the alarm switch or else cut the wires leading to it. There are some types of alarms that cannot be defeated by cutting wires. Once the alarm switch is closed, a module containing solid-state circuitry is activated, making the alarm independent of the switch. Also, some alarm switches are wired in series with the ignition switch. But the car will not start unless the alarm switch is turned on, creating precisely the condition the thief does not want. The car owner, however, can defeat the alarm switch, and so can start the car without sounding the alarm.

AUTO ALARM SYSTEMS

A large variety of auto alarm systems is available having various degrees of sophistication. It would be impossible to describe them all. The few that are mentioned here are just to give you some indication of the different ways in which you can protect your car.

Protecto Lite

With the help of Protecto Lite, you can turn your car lights on to light your way from as much as 300 feet away. In case of street attack, just press the button of the hand-held remote control unit and your car horn will blow for 3 minutes, long enough to frighten any attacker away. Pressing the button the first time turns on the interior lights and headlights. Pressing the control button again will cause the car horn to become activated for 3 minutes. To stop the alarm, just turn on the ignition. The system automatically arms itself when the ignition is turned off. See Fig. 6-4.

Fig. 6-4.

Protecto Lite turns on car interior lights and headlights for 3 minutes, can also activate car horn from up to 300 feet away. (Courtesy A.C. Custom Electronics)

Steal Stopper

The Steal Stopper is available in four different security arrangements. One is an ignition lockout system in which the car will not start. The second is an ignition lockout with the car horn used as the alarm. The third is an ignition lockout with a siren and motion detector that senses jacking up or towing of the vehicle. The fourth is an alarm unit only and is equipped with a siren and motion detector.

In some of these systems a 12-digit keyboard (Fig. 6-5) is used. You can enter a secret 4-digit code that you preselect from 11,860 combinations. Under 2 inches wide, the keyboard protects your car even if you forget your keys in the ignition, because the car will not start until you enter your 4-digit code.

The digital keyboard mounts onto the dash while its control module, put under the hood, connects to the battery and ignition. The keyboard cannot be hot wired or master keyed from inside the vehicle.

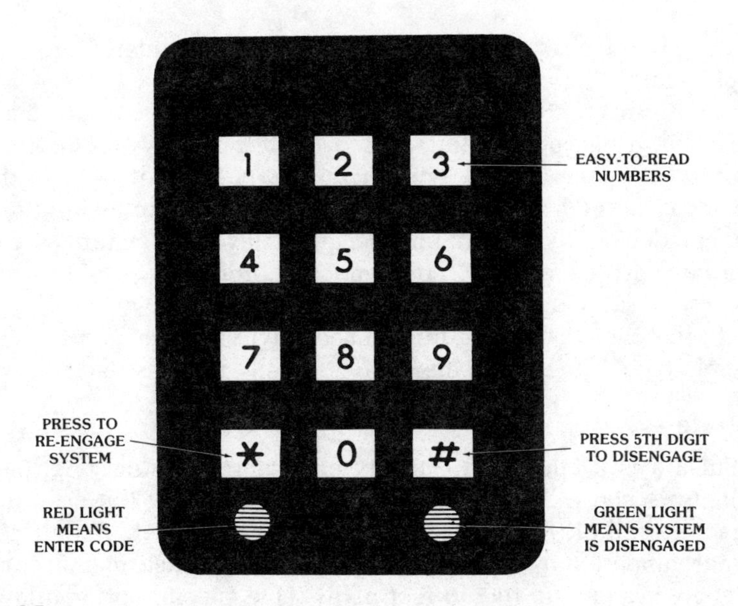

Fig. 6-5.

Digital keyboard for Steal Stopper auto security system. (Courtesy A.C. Custom Electronics, Inc.)

To use, simply turn your ignition key on and press four digits in sequence to start your vehicle. By pressing a fifth digit you can temporarily disengage the system for attendant parking, car wash, etc.

The Steal Stopper has two separate indicator lights, with a red light meaning enter code and a green light signifying system disengaged. The unit automatically resets itself each time the ignition key is turned off.

Scat III

The purpose of Scat III is to prevent a break-in before it occurs. The unit listens for the sounds of attempted breaking and entry with a sensitive, concealed miniature microphone. Every sound heard by the Scat III is processed through a sound discriminator computer programmed to analyze and recognize the noise an intruder makes when he tampers with your car. Before the thief has a chance to add you to his list of victims, Scat III sounds a siren and turns on lights. The alarm reacts before a door is opened. The purpose of Scat III is to fight back against the professional car thief, but in so doing will also protect against the other types of thieves, the parts stealer, the joyrider, and the opportunist.

If, with this system, you should happen to leave your car doors unlocked, a microcomputer takes over. It monitors all electrical activity in your car and sets off an alarm when it detects the dome, trunk, or hood courtesy lights. At the time of installation, you can decide if you want an 8-second delay or an instant alarm. You can also have prewired connections for protecting gas-cap covers, pick-up truck tool chests, van doors, or wherever you might feel the need for additional security. The siren is a 10-watt warble type. The system also automatically re-arms itself.

Pulsafe

Pulsafe is a combined security and control system (Fig. 6-6). Pulsafe is controlled by a lightweight transmitter the size of a cigarette lighter, fitting easily into a purse or pocket. With it, the owner, and that person alone, can arm and disarm the car's security system, up to 250 feet away. One special option allows the owner to lock and unlock all the car doors automatically when he arms and disarms the system, all by remote control at the touch of a button.

The transmitter sends a coded radio signal to the system hidden inside the vehicle. Since there are 140,000,000 different code combinations, it is virtually impossible to break the code for any single vehicle.

The system detects either forcible or stolen key entry. Once upset, the system blasts the horn and flashes the lights. The alarm continues for 60 seconds and if the thief persists in trying to enter the car, the cycle is repeated as long as the tampering continues. Even if a persistent thief were able to get into the car, he would not be able to start it because this security system disables the fuel or ignition system. Not even "hot-wiring" will work. The system also offers an optional standby battery and siren to thwart the thief who might disable the car battery, thereby shutting down the car's security system.

The system confirms arming and disarming commands with visual and audible signals and warns the owner as he enters the car whether the vehicle has been tampered with by sounding one warning beep of the horn when any door is opened. If the owner forgets where he has parked his vehicle in a large parking lot, the transmitter in the unit can locate the car up to 250 feet away. The

Fig. 6-6.

Pulsafe car security and control system. (Courtesy TMX Systems, Inc.)

system responds to the owner's coded transmitter signals by honking the horn and flashing the lights.

If the owner finds himself or herself unexpectedly in danger of attack while inside the car, a hidden personal protection panic button can immediately put the system in its alarm mode. The unit also warns the owner with an audible signal if the lights have been left on or if a door is open. As shown in Fig. 6-7, the security system can be controlled even through a brick wall.

Protection Areas

A sensor is a device that responds to a change. The change

could be one of motion such as going from a complete stop to movement, or it could be vibration, such as that supplied by a car that is being driven. A sensor can be made to respond to lights, or heat, to the opening of a door, to pressure such as that supplied by a person sitting down in a car. Fig. 6-8 shows the various areas or points in a car that can be used with sensors that will trigger a car alarm system.

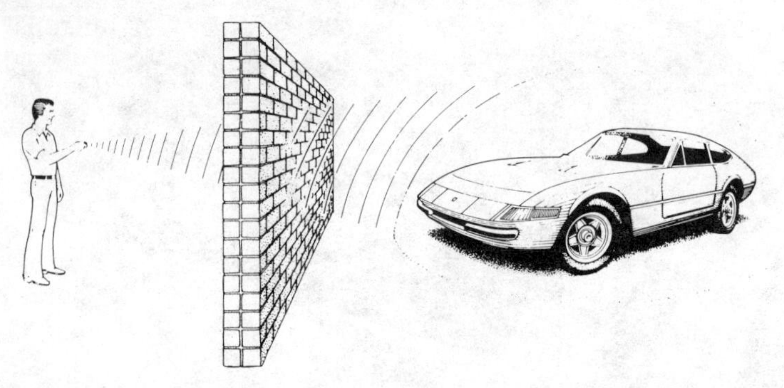

Fig. 6-7.

Car security system is controlled by radio wave that can penetrate a brick wall. (Courtesy TMX Systems, Inc.)

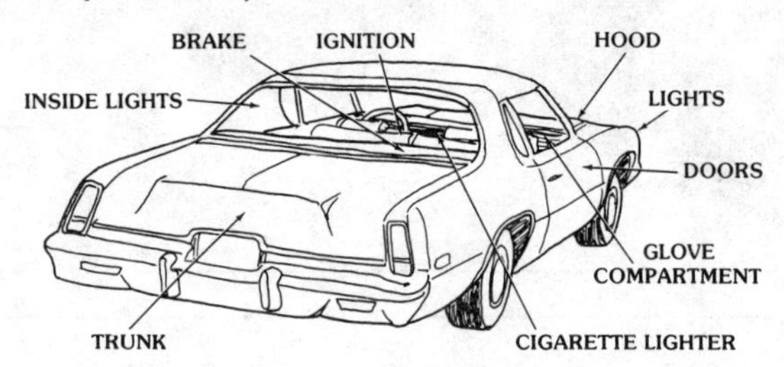

Fig. 6-8.

Various areas or points on a car are suitable for a change-of-state sensor.

A pressure sensor can be positioned directly beneath the driver's seat, as indicated in Fig. 6-9. A flexible switch beneath the seat turns the alarm on. The switch can also be arranged so that the pressure of the thief's body turns the strip switch off. If the strip switch is in series with the ignition switch, the car will not start.

Arrangement of an Auto Alarm System

Fig. 6-10 shows one possible arrangement for an auto alarm

system. The secret switch can be located somewhere under the dash. Turning on the switch when leaving the car sets the alarm system to its ON position and may also disable the ignition. The module contains a circuit for delaying the alarm turn-on to give the car owner time to close the car door. The module also turns the alarm ON if a thief tries to gain entry at various points or touches various parts of the inside of the car. It also recycles the alarm, turning it OFF after a certain amount of time and then ON again.

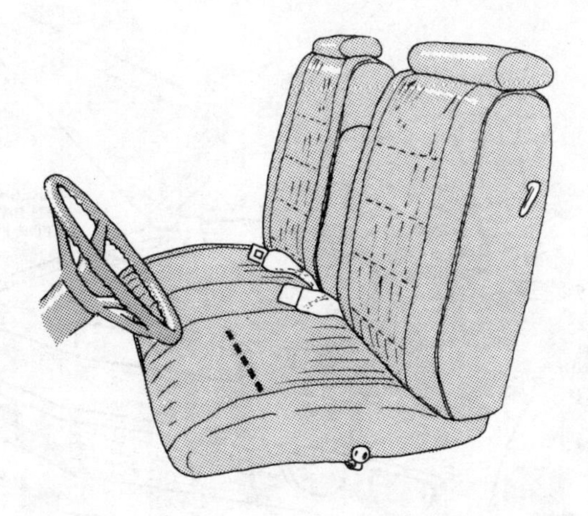

Fig. 6-9.

Unexpected flexible switch beneath seat covering sets off alarm. (Courtesy Tapeswitch Corp. of America)

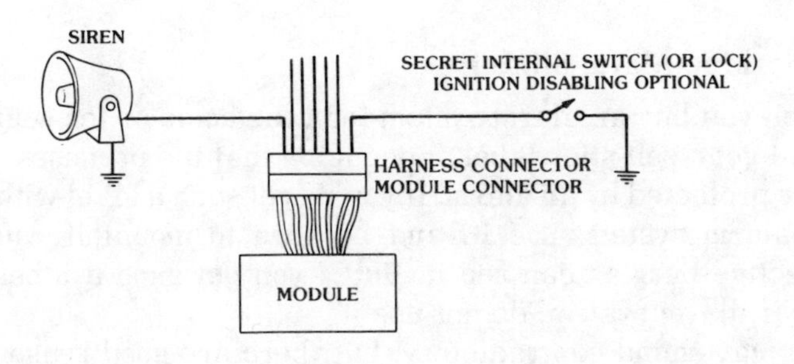

Fig. 6-10.

Typical arrangement for auto alarm system; secret switch can be located somewhere under the dashboard.

Auto Alarm Installation

Fig. 6-11 is a diagram of an auto alarm installation. The unit is set to its ON or protective mode by a key switch mounted on a fender. The car battery supplies the power for the alarm. The module contains circuits for recycling the alarm, turning it off and on for predetermined amounts of time. The hood switch turns the system ON when the hood is lifted just an inch or two.

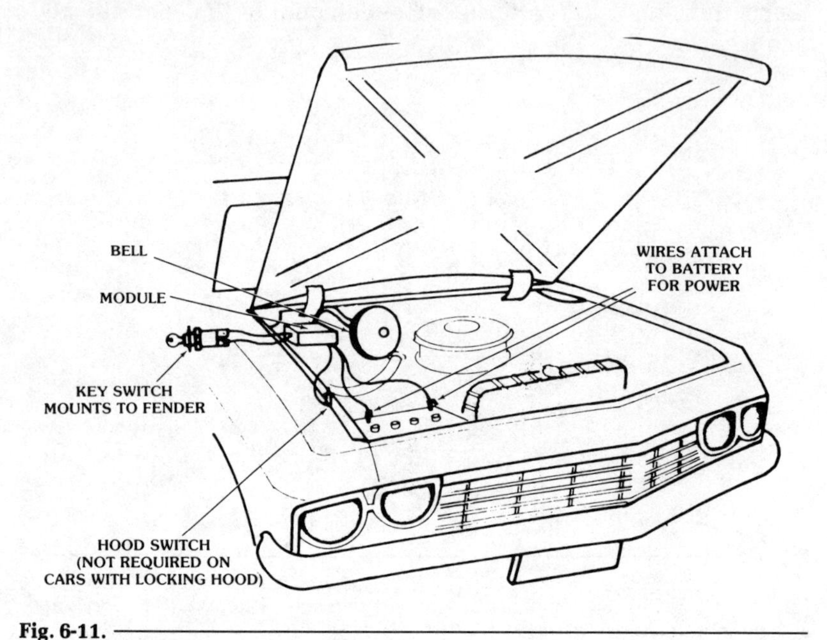

Fig. 6-11.

Auto-alarm installation diagram.

The Warning Alarm Sticker

When you buy an alarm system for your home or for your car, you will get a self-stick label announcing that the premises or the car are protected by an alarm. If you do get such a label with your home alarm system, use it, and be sure to mount it where a prospective burglar can see it. But if you get such a label with your car alarm system, do *not* use it.

This may sound contradictory, but there are good reasons for this kind of action. With a home alarm system the burglar has no way of getting at the alarm. He doesn't know what kind it is, doesn't know if you mean the door is protected, the windows, or all of the house. He doesn't know if you have an ultrasonic alarm, a radar alarm, whether you are area or object protected. And because he doesn't know, it is quite likely he will move on to a less security-minded prospect.

Now consider your car. You have a sticker on it saying the car is alarm protected. There's no question about the kind of system it is and the thief knows it. He can get under your car, quickly locate the wiring to the alarm, cut the wires, and that's the end of your protection. With a car alarm system the best thing to do is to keep the thief guessing. Do you have an alarm or don't you? This is one of the thoughts in his mind, and you can be sure he is prepared to make a quick getaway the moment an alarm is heard.

On many car alarm systems, the lock for the alarm is mounted in an easy-to-get-at position. There is a reason for this. It is convenient for the car owner to be able to set the alarm with a key, and so the alarm lock is mounted in some conspicuous spot. This is equivalent to notifying the thief that the car is alarm protected, and performs the same function as the alarm sticker or label. Knowing the car is equipped with an alarm, the thief can take steps to disable it. Further, by positioning the alarm lock where it can be readily seen, you have, in effect, also told the thief where he must reach in to cut to disable the alarm system.

There is still one other serious disadvantage to the alarm label and the visible alarm lock. Many thieves reason that only the owner of a car containing valuables would go to such trouble or expense. Of course, they could be wrong but this won't stop them from making an attempt at your car. If they can disable your alarm system, having been alerted by you that such an alarm exists, then your car becomes more desirable than some other vehicle that may or may not have an alarm.

If you do have an alarm system in your car, don't expect it to do the impossible. An alarm is no guarantee that your car won't be stolen. All it does is assure you that there may be some noise from your car siren. But if you have your car parked at some remote corner of a parking lot, and some of them are tremendous, the wail of your siren will be ignored. It's your car, your property, and no one is going to run a great distance across some car-parking area to get involved in a possible clash with a thief. If you are going to park, try to get as close to the entrance as possible. This is where you will find the parking attendants. Or else select a parking lot that is small. And if you park out in the street, try to park near a house or store with some activity. Park in a deserted street, adjacent to a vacant building, and all that will happen is that your siren will scream until the thief can get at the connecting wires. Car alarms have limitations and are valuable only if you recognize these limitations and work within them.

GAS DISCHARGE PROTECTOR

Various esoteric schemes have either been proposed or tried for protecting the car against theft. One of these is a cylinder containing a gas under compression which is triggered for release on breaking and entering. Such an arrangement has doubtful value. The thief may not respond quickly to the gas, or it may cause him to have an accident for which you could be held responsible. It could result in the death of the thief, raising the question as to whether you have the right to kill in defense of property.

TIME DELAY PROTECTOR

Another possible car protection device is a time-delay fuse connected to the spark coil. The thief gets into your car, starts the ignition and is able to drive a short distance, usually less than a block and then the time delay fuse opens. The car stops and there is the thief with a car that will not move, away from the safe and quiet area where you had parked your car. He is now out in open traffic where at any moment a policeman may come along to find out why the car is blocking traffic with, of course, the usual request for license and/or registration.

The system has the benefit of being simple, perhaps too much so. There is nothing to prevent the thief from abandoning your car if he finds it will longer respond. So there is your car out in the middle of the street or a road, with the doors wide open, but with you as fully responsible as if you had driven the car there yourself.

There is also the problem of disabling the time-delay fuse. Presumably this could be done by a hidden switch somewhere under the dash which would short the fuse, making it inoperative when you wished it to be so.

DELAYED ALARM

Another suggested method is to have an alarm activated, but

time delayed so the alarm doesn't go off when the thief first enters your car, but after it has moved out into traffic. All you have achieved though is to have put the thief in an awkward position. He does the only thing he can do under these circumstances. He runs. But your car, blasting away, and doors wide open, is out in the street. The fact that your car was stolen does not relieve you of responsibility.

TEMPORARY OR PERMANENT THEFT?

If your car is stolen, whether or not you will get it back depends in part on the kind of thief who made away with it. About half of all car thefts are attributed to juveniles who want a car, want it now, and don't care whose car it is. These cars are usually recovered, since the thief abandons the car as soon as it runs out of gas. Whether or not the car will be damaged is quite another matter. Some juveniles are much better thieves than they are drivers. Since part of their enjoyment is in risk taking, the odds are against your getting the car back without some damage.

The easiest car to steal is the convertible: a simple cut through the soft fabric and the thief has his hand right inside the car. All it takes is a sharp razor mounted in a holder, such as those designed for using the razor as a scraper. The next easiest is the older style of car that has a small side or vent window. They have rubber seals. The burglar inserts a screwdriver between the seal and the frame and can often lift the window enough to get at the side latch. If not, he uses a wire coat hanger and works it in until it catches on the latch.

DON'T BUY A STOLEN CAR

It can be quite easy to buy a stolen car, and it can be quite tempting, particularly if the price is low. Sometimes a stolen car will be advertised and the seller will supply a number of reasons as to why the deal must be an all cash arrangement. No matter how valid they may sound, your suspicions should be aroused and you should back away just as fast as you can.

If you buy a car that has been stolen, and if it is traced to you by the police, you will not only lose the car but may be subject to prosecution as a receiver of stolen goods. If you buy your car from a dealer, and it subsequently develops that the car was stolen, you at least have a chance for a refund. If the dealer has been in business for a number of years at the same location, if he hasn't acquired a reputation for dealing in hot cars, the resonsibility for merchandising a stolen car may well be his. And if the car has a lien on it, and he does not so inform you, you have a case for collecting from the dealer.

THE GAS STATION RIPOFF

When you use your car for traveling, especially through the various states, just one look at your license plate informs the gas station attendant you are a transient and are not likely to drive back any distance to protest a rip-off. Most gas stations are reliable, but there are some that always welcome a quick unearned dollar from out-of-towners. So when driving out of town you will need to change some of your gasoline buying habits.

It is often difficult to see the indicators on a gas pump that inform you of the amount of gasoline and the total cost if you insist on remaining in the driver's seat. Instead, get out of your car and keep an eye on the pump. If you have an opportunity to use a self-service pump, do so. Do not leave your car unattended to use the restroom facilities. Instead, wait until you have received and paid for your gasoline, drive away from the pump, and park in a convenient area near the restroom.

If you think you need an oil change, ask the attendant to let you see the dipstick after he has checked for oil. Most attendants will do this as a matter of course. However, you may come across a service station attendant who will put the dipstick in only part of the way, will tell you that you need additional oil, and will use an empty container and go through all the motions of supplying your car with more oil.

CAR STRIPPING

A car does not need to be abandoned to be stripped—it must just look that way. If your car becomes inoperative while driving along a highway, do not leave your car, or at least let one or more (preferably more) passengers remain with the car while you go for help. Above all, do not leave your car alone overnight on any road. Vandals will be sure to take over where the strippers leave off.

THE UNMARKED SQUAD CAR

The fact that you are driving in your car, with the windows closed, may give you an unjustified sense of security. What do you do if another vehicle pulls up alongside you and signals you to stop? You think it is an unmarked police car but you are not sure.

In a situation such as this, even police departments offer conflicting suggestions as to the best course of action to follow. Some officials advise that you should continue moving until you reach a safe place to stop. This would be under a street light or somewhere out in the open. You should then wait until the unmarked car parallels yours with the occupants offering some identification.

Unless the unmarked car has an emergency light, a siren, and the driver can show a police shield, the best thing to do is to keep your doors locked, your windows closed and not permit your vehicle to get blocked in.

If it is a police officer who has stopped you, he will not just sit in his car waiting for you to exit from yours. Instead, the police officer will get out of his car and will show his identification. The police officer may or may not be wearing his uniform. If he is a plainclothesman, he will show his identification.

One way in which you can recognize an unmarked squad car is by the behavior of the occupants. You will be directed to a well lighted spot and out of the way of traffic. You have the right to demand identification. Further, you must be allowed to examine the identification to make sure that it is legitimate. Be suspicious if the person who has stopped you just waves a card in front of you and then just as quickly puts it away.

If your car is stopped, be sure to keep your hands visible and instruct your passengers to do so as well. Your car may have been stopped because it resembles one used in a holdup. Remember, also, that if the unmarked car is legitimate it is probably radio equipped and can call for a regular squad car for help. So if you are not satisfied, just keep driving.

WHEN TO PROTECT YOURSELF

Many car owners install some form of security system after they have been robbed. Someone who loses a battery, or wheels, or a tape unit, has the idea of security forcibly brought home. But after the anger subsides and the security system is installed the usual let-it-go attitude prevails once again. Car doors are left unlocked, keys remain in the ignition system, and the entire pattern of carelessness and theft is repeated. No lock is worth the room it takes up unless it is used. No alarm system is worth carrying around in your car unless it is in working condition, ready for action. Thieves may be lazy but the one great factor working in their favor is that they are no more so than the ordinary individual.

SUPPOSE YOUR CAR IS MISSING?

The first thing to do is to make really sure your car is missing. Some department-store shopping malls are tremendous and, unfortunately, may not be equipped with poles carrying identifying letters and numbers. In that case, locating your car can be a problem. If you cannot find your car and then report it as stolen, only to locate it later, be sure to report it as returned or found, otherwise it will remain on police records as a stolen vehicle. You may be embarrassed at some later date if you are stopped for a violation or when you renew your registration.

If your car is stolen, report it as quickly as possible to your bureau of motor vehicles. Also telephone your insurance agent or notify your insurance company.

If your car is found and returned to you, once again inform your insurance agent and/or your insurance company. It would be unrealistic to expect your car to be found in the same condition in which you left it. If you car was stolen and just taken for a ride, it may be found abandoned by the police who may then tow it away. You will then need to establish proof of ownership. What you can collect from your insurance company depends on the company and the terms of your policy. Don't expect to make a profit.

YOUR OTHER VEHICLES

The trouble with protecting motorcycles is that practically none of the security features used for automobiles can be applied to the problem. A pair of thieves, working as a team, can literally pick up a motorcycle, put it in a van or truck, and drive off with it. And if the motorbike has an alarm, so what? Covered with a blanket in the van, completely enclosed by the walls of the van, the muffled sound it will emit won't attract attention.

How to Protect Your Motorcycle

About the only solution is the use of a chain and a lock. This does not mean the ordinary bicycle chain that can be snipped. It does mean the heaviest, most solidly forged chain you can get and a tough, case-hardened padlock. And the post you wrap the chain around must be equally tough. Don't pick a small tree and think you are safe. A thief can saw through that rather quickly.

If you park your motorcycle in your own garage, at least make sure the doors cannot be forced open. If you have any doubts, chain-lock your motorcycle just as though you had parked it out in the street.

Keep a record of your motorcycle identification number, the manufacturer, model number, and year. Use an electric pencil to mark your identification on it in at least two places, preferably spots that aren't readily noticeable. Keep your bill of sale to prove the motorcycle is yours. And make sure your theft-insurance premiums are paid.

Many of the precautionary rules for cars apply to motorcycles. Don't park on deserted streets, or in high-crime areas, or in the remote end of a parking lot. If possible, don't park at all, and if you do, at least try to put the motorcycle where you can keep an eye on it. Yes, this does destroy the fun of owning a motorcycle, but theft destroys the fun of everything else as well.

How to Protect Your Moped

Like the motorcycle, the moped is a two-wheel vehicle. It is lighter and less expensive than the motorcycle and in some states the highways on which it is allowed to travel are restricted.

The security suggestions for motorcycles apply equally to mopeds. However, the moped is a much lighter vehicle and so, if it is chain attached to a post, can be lifted until the chain clears that post.

When chaining the moped, make sure that the chain passes through some part of the body of the moped. Some moped owners chain the front wheel, but that wheel is not only easily removable, it is easily replaced. Be sure to put an identifying number somewhere on the moped with an electric pencil. Don't just write your name with a felt pen, because such writing is readily removed. Keep a record of the model number, type, serial number, if any, of the moped together with your purchase receipt. Check with your insurance company to learn if you can add theft and accident insurance to your existing motorcycle and/or car insurance.

The safest place for a moped is in a locked garage and it is even safer if you use a chain and padlock on the moped while it is in the garage.

Take a color photo of your moped or make a color videotape of the unit. A color photo is better in case you need to identify the moped for the police and/or your insurance company. You may need several photos, so keep the negative or make several photos at the same time. On the back of the photo, using a felt-tip pen, write all pertinent data, including any distinguishing characteristics, make and model number, the price you paid, and the date.

How to Protect Your Bicycle

Bicycles are now enjoying renewed popularity. As a result, bicycle sales have been booming, and many new styles and models have been introduced. At one time bicycles were relatively inexpensive, but with the addition of special bicycle features, such as a multispeed gear shift and hand brakes, some have moved up to several hundred dollars or more, making them particularly tempting targets. Bicycle thefts have kept pace with the burgeoning bicycle market.

To leave a bicycle unattended and unlocked is simply an invitation to theft. All the thief has to do is mount it and ride away.

The most common method of securing a bicycle is to use a key-type padlock and a length of chain, ordinarily about three feet long. The chain is often enclosed in a plastic sleeve to protect the bicycle paint finish against chain scratches. This system supplies a moderate amount of security, but with some negative factors.

The first of these is the device to which you fasten the bicycle chain. If this is a pole, for example, and the pole is short enough, all the thief has to do is to lift the bicycle and chain over the pole. It is true the chain will still be attached to the bicycle, but then the thief can remove the bicycle to some remote area where the chain can be hacksawed off.

A thief can hacksaw his way through most chains, so if you are going to use a chain, get the strongest and toughest one you can buy. Chains come in various degrees of thickness and hardness, so the amount of security you will have will depend on these qualities. The lock is also important. Just because a lock closes doesn't mean it can't be opened by being smacked with a hammer. The sturdier the lock, the more it will cost.

Percentagewise, security for a bicycle is more expensive than for most other forms of transportation. It is possible to buy a bicycle for 80 dollars and to spend about 16 dollars for protection. This is 20 percent of the orignal cost. Comparably, for a car costing 4,000 dollars, this would mean spending 800 dollars.

Whether an attempt will be made to steal a bicycle depends also on where you park it. In a high-crime area, any bicycle left parked for one day will certainly be subjected to one or more "theft tests." This means that one or more thieves will try to steal the bicycle. Whether they will be able to do so or not depends entirely on how the bicycle is protected. Not even with an automobile do you have such assurance that thieves will try to get away with your property.

At one time, bicycle thieves relied on heavy-duty, specially made hacksaw blades to cut through bicycle chains. Today they have the services of a chain-link cutter that does the job faster and easier. Whether the chain link cutter will be successful or not depends on the size of the link cutter and the size and strength of the chain.

To minimize the possibility of bicycle theft, manufacturers now offer various chain-and-lock combinations, designed to resist sawing or cutting.

Some bicycles are made with a removable front-wheel feature. This doesn't relieve the owner of the necessity for chaining the bicycle. After doing so, he removes the front wheel and takes it with him. There is no question that this is an additional deterrent, and it eliminates the possibility of the thief riding away with the bicycle. But it is a nuisance, and it does mean an assembly and disassembly job every time the bicycle is parked.

Bicycles aren't licensed in most towns and cities, so there is no bicycle identification number you can use to prove a particular stolen bicycle is yours. However, with an electric pencil, you can etch your initials somewhere into the frame of the bicycle in an inconspicuous place. Also etch your name or initials into the underportion of the bicycle seat. The thief can replace the seat, of course, but quite often he will not bother.

When you buy a bicycle, be sure to keep the purchase receipt. This will not only help prove ownership, but you may need it to show how much you paid for the bicycle and when you bought it in the event your insurance covers bicycle theft.

BICYCLE ANTITHEFT CHECK LIST

1. Just because your bicycle is in your own driveway doesn't mean it cannot be stolen. Lock the bicycle just as though you were parking it out in the street.

2. Get the best padlock and chain you can buy.

3. Try to avoid parking your bicycle in high-crime areas.

4. Try to avoid parking your bicycle in deserted, poorly lit areas. Overnight street parking is a sure invitation to theft.

5. Don't simply snap the lock shut when you park. Check it by pulling on it. A lock will sometimes look closed when it is not.

6. A chain and padlock are no more secure than the pole to which they are attached. If you fasten your bicycle to a wooden post, for example, it will require no great effort on the part of the thief to cut or smash through the post.

7. Make sure your bicycle is electrically marked with your initials, preferably in two places. Also be sure your name or initials are marked on the inside of the seat.

8. Keep your purchase receipt. Also memorize the name of the manufacturer of the bicycle. Be able to identify the bicycle not only by its color, but by its features as well. You should also know the bicycle model number, if it has one. You can get all this information from the descriptive catalog sheet furnished when you buy the bicycle, or it may be supplied as part of a manual. If not, then be sure to get this information from the store where you buy the bicycle. You may have to supply this information to the police and the more specific and detailed your data, the better the chances of recovery.

9. Don't lend your bicycle to friends, neighbors, or strangers without realizing that the full responsibility remains yours. You are simply lending a bicycle; that is all.

10. If you keep your bicycle parked in your garage, or in a basement, lock your bicycle just as though you were parking it out on the street. Garage doors are frequently left open and an opportunistic thief can walk in and ride out with your bicycle in a matter of seconds.

Chapter 7

Travel Security

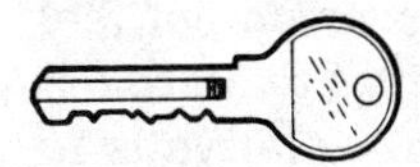

The possibility of being robbed is a form of pressure. But unlike other forms of pressure, such as job pressure or the pressure of ordinary daily living, the pressure of potential robbery with you as its victim is unrelenting and unceasing. It is this way because you never know when you will become the target. The old concept of theft by night is now a myth. Daytime robberies are now commonplace, and, as a matter of fact, the element of surprise now plays a large part in some robberies. As an example, you may be walking down the corridor of your hotel toward your room and as you reach the door a gun, seemingly from out of nowhere, will be in your back, with a disembodied voice suggesting you do nothing foolish. Within a moment, you and the thief are alone in the room, where he has an opportunity to rob you at his leisure.

This is an opportunisitic type of robbery—that is, you became a victim because all the circumstances were right. You were alone, there was no one else in the corridor, and you were about to open the door to your room; at least one of your hands was occupied, making you defenseless to that extent, and you may have had a purse or package in the other hand. Preventing this type of robbery is almost impossible unless you can always make an arrangement to have someone with you at all times.

HOW MUCH CASH SHOULD YOU CARRY?

However, if there is no alternative to robbery, you can at least plan ahead to minimize its effects. One method is never to carry any more cash with you than is absolutely essential. If you carry money, try to arrange to have as much of it as possible in travelers' checks. They are low in cost, readily obtainable in banks, and you can get a refund if you are robbed. Of course, you should keep some currency in your wallet or purse, but this should not be more than necessary to cover daily requirements.

Money Belt and Money Pouch

For some individuals, however, cash in a wallet or purse is a form of psychological security blanket. Such persons do not feel comfortable unless they have some assurance that they can reach cash and touch it in a matter of moments. The alternative is to carry a money belt (for the men) and a money pouch (for either men or women). The money belt looks like the usual sort of belt, but has a compartment in which a small amount of bills can be cached. The money pouch looks something like a belt, but can be used to carry a larger sum. Unlike the belt it is worn beneath the clothing and, aside from some minor discomfort, is a good money-security measure.

Hidden Money Compartment

For the men, a wallet having a hidden money compartment is also available and this can be used to secrete cash. There is always the possibility that the thief will take the wallet as well as its contents, but then again he may not and so you have at least some chance that you will not lose all your money. In some cases, the thief may compel his victims to strip, and so the presence of the money pouch will immediately become evident. But he may not require this action and so the use of the money pouch will also keep the loss from becoming total.

The sum and substance of all this is that you have no guarantee. The behavior of a thief is unpredictable. Some will resort to violence if they feel that their time and efforts have not been adequately rewarded. A thief who needs fifty dollars for a fix will not willingly settle for ten. He is well aware that a second robbery for the additional money he must have will not only take time but will increase his risk proportionately.

Do Not Carry Large Bills

Aside from the money belt, the money pouch, and travelers' checks, there are a few more precautions you can take. Avoid the habit of carrying large bills. Ten-dollar denominations are much better than twenties and twenties are certainly better than fifty- or hundred-dollar bills. The sight of a large roll of bills can attract attention. In that case the thief will follow you until you give him the opportunity to make his move in relative safety.

To avoid making an ostentatious show of money, you might try keeping most of your cash in a money belt or pouch, but if these don't appeal to you, you might consider the two-wallet technique. With this method you keep most of your money in one wallet, with that wallet carefully secured in an inside pocket (never a hip pocket). This wallet should be zippered into position and the pocket should be deep enough to cover the wallet completely. You can then keep your spare currency, enough to cover your daily needs, in the other wallet.

WOMEN AS TARGETS

Many thieves prefer women as targets of opportunity for a number of reasons. One of these is that it takes a woman much longer to reach into her pocketbook for her money purse than it does for a man to reach for his wallet. A man can have his wallet out and back again in a fraction of the time it usually takes a woman to fish around in her pocketbook for her purse, thus giving the thief a longer opportunity to snatch the purse.

There is also a big difference, possibly psychological, in the way men and women pay their bills. If people are lined up ready to pay a cashier, most of the men will have their money in their hands before reaching the cashier. A woman will usually wait until she is directly in front of the cashier before opening her purse. A thief, aware of these male-female differences, cannot predict when a man will take out his wallet. All he knows is that it will be sometime while he is waiting in line. But, under some pretense or other, he can wait directly in the vicinity of the cashier, and assess rather well just how much cash a woman has available.

Another reason a thief may prefer a woman as a target, other than the obvious one that the woman is usually weaker, is that a woman's purse may be much more accessible than a man's wallet. The woman's purse is in her pocketbook, but that pocketbook is external to the woman's clothing. It is easier to jostle, to move it, and to do things with it. A man's wallet isn't external to the man's clothing, but is actually imbedded in layers of it. This doesn't mean a man's wallet cannot be removed without detection. It happens regularly, but it does require the effort of a professional. For the amateur—and there are many more of those than there are professionals—the woman's pocketbook is a more tempting and more available target of opportunity.

There is still one more reason why women are becoming more likely victims and the fault—if blame is to be assigned—lies with the men. In many marriages women not only work as housekeepers and mothers but also perform the functions of a treasurer. In many households a wife is expected to balance the budget, handle all finances, maintain the checking account in proper shape, and to pay during a shopping expedition. This means she carries the money—not all of it, but quite often most of it, particularly if the couple are traveling. All the more reason, then, for the woman to become the victim.

EYEBALL TO EYEBALL CONFRONTATION

For the victim a robbery can be, and often is, a traumatic experience. But, aside from the fact that a robbery is the involuntary transfer of some of your wealth to a stranger, a robbery in which the victim and the thief are fairly close to each other is a form of communication—not a pleasant form, but communication nevertheless.

To the thief, the victim is nameless and faceless, not a human being, but an object barring the way to the money he wants. If he resorts to violence, he will do so because he isn't taking action against another human being. But you can become more than just an object if you look into the eyes of the thief. This eyeball-to-eyeball confrontation is a form of communication in which your relationship, not as thief and victim, but as a pair of human beings, is established. When this happens, the thief is less likely to resort to violence. It doesn't mean the robbery will stop, but it does mean there is less likelihood you will be hurt. Yes, there are exceptions, as there are to almost everything else. Many victims are so terrified that they concentrate on the weapon instead of the person holding it. Of course, it is easy to give advice, and it is certainly more difficult to remember what to do to prevent a robbery and how to act during one, but at least you will have been forewarned, and that may help.

JUDO AND KARATE

Judo and karate are two of the more common forms of self-defense. Many individuals take courses in these activites as protection in case of attack. There is nothing wrong with that, but generally these efforts are undertaken in a burst of enthusiasm that dies out after two or three lessons. A small number of lessons will not only not make you an expert, but may give you a wholly unwarranted feeling of self-confidence. It takes a lot of effort, self-discipline, and continuous practice to become a judo or karate expert, and if you have done this, then you are indeed in a position to repel an attack. A determined thief with a knife, gun, or other potentially lethal weapon is usually more than a match for a beginner in judo. Practice judo before you are robbed, not during.

HOTEL THEFT

You can be robbed in your hotel room even though you have the door locked. Since the same key to a hotel (or motel) room is used by so many people—including the fact that there is always a master key around somewhere—it shouldn't be too surprising that having a duplicate key made is no great problem.

Your hotel or motel room, then, is not safe while you are away simply because keys are so readily accessible. The room may be safe, when occupied, if you do more than just close the door. Such rooms are generally equipped with deadlocks that can be adjusted only from inside the room. Don't assume you will close the deadlock at the time you retire. You may or may not. Lock the deadlock as soon as you close the door. In that way you will be sure to avoid unwelcome visitors.

A special traveler's lock is available if you want hotel- or motel-room protection while you are away.

KEYED TRAVEL LOCK

The keyed portable travel lock, Fig. 7-1, can be used to double lock hotel or motel doors, secure valuables in drawers or closets, lock filing cabinets, lock double doors, secure medicine chests or lockers. In use, the locking bar is installed. A locking plate is then slipped on the locking bar and the lock housing is pushed on the bar to secure the plate and the item being locked. The locking

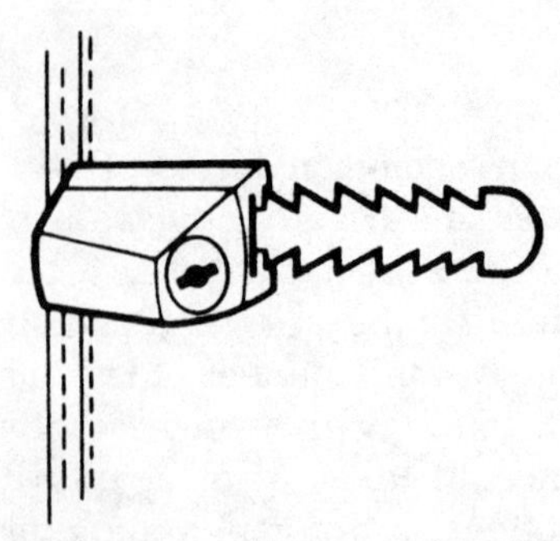

Keyed travel lock. (Courtesy Taylor Lock Co.)

Fig. 7-1.

plate has a T cutout, enabling it to be put on the locking bar horizontally or vertically. This doesn't mean the thief can't overcome the lock. He may be able to do so with enough time, but generally the sight of such a lock is enough to encourage him to move along to a door that doesn't present such problems. After all, there are numerous rooms available and there is no reason why he should spend an unnecessary amount of time on one that poses a problem.

If you have valuables with you in a hotel or motel room, it may be better to deposit these in a safe-deposit box provided by the management. You should also learn of the extent of responsibility for theft assumed by the management. And, you should also have traveler's insurance on all valuable items you carry with you. Check with your insurance agent on the extent of your responsibility—that is, whether you are required by your policy to store your valuables in a hotel safe during night hours.

LOSING YOUR LUGGAGE

The suitcases you carry aren't made for you alone, even though you may have personally selected them for their style, shape, and color. First, make sure that the luggage you buy comes equipped with at least one lock, preferably two. Make sure these are locked, not simply closed shut, when you travel. Attach a name tag to each piece of luggage, so that if you and the luggage become separated, there is a chance the finder of the luggage will find you also.

But this isn't enough. Put a self-stick strip of colored plastic across some prominent area of your luggage. This will immediately identify it as yours. It will make it much easier for you to locate your luggage when it is mixed in with a large number of other pieces of luggage, and it allows for easier identification if your luggage should happen to stray. If your luggage does become lost, complain immediately to the transportation service: bus, cab, train, ship, or plane. Most transportation companies assume very limited liability for luggage and its contents. Depending on what you carry it may (or may not) be worthwhile for you to carry supplementary insurance.

If you must turn your luggage over to someone else, as in the case of a plane, be sure to get a baggage receipt and do not release this receipt except in exchange for your luggage. This receipt will not only document the luggage as yours, but fixes responsibility on the carrier, and should help in the location of your luggage, assuming, of course, that it hasn't been stolen.

The best thing you can do for your luggage is to keep an eye on it. As soon as the luggage is returned to you by your transportation service, you are once again fully responsible. Never leave your luggage unattended. Thus, putting your luggage outside a men's or ladies' room while you go inside for just a few minutes is inviting theft. And don't ask strangers to keep an eye on your luggage for you while you go off to make a phone call. The stranger may do more than you have asked. If you do make a phone call, squeeze the luggage inside the phone booth with you, even if it means you will be somewhat crowded. In the brief moment you may turn your head aside while inside the booth your luggage, carefully placed outside, can literally vanish.

Small-change thieves—those who are satisfied to make just a few dollars—often make railway, plane, and bus terminals their places of business. They seem to favor bus lines, though. Do not use the services of anyone offering to act as a porter, even though the offer may come from a young, pleasant-looking boy who is apparently just trying to make an honest dollar. He may be young—many bus-terminal thieves are—but honest he is not. Keep a strong grip on your bags and surrender them only to a bona-fide porter. He will be wearing a uniform and his cap will probably have the word "porter" written on it.

But no matter who carries your bags, do not let them out of your sight. Bags cannot only be stolen, but they can become lost or strayed. Tracing your luggage can be time-consuming, aggravating, and extremely inconvenient. When paying for luggage-handling service, don't reach for your wallet. Keep enough loose change in your pants pocket to pay for this service. This will enable you to pay without revealing the location and contents of your wallet.

YOUR CREDIT CARDS

Most persons who travel keep their credit cards in the same wallet with their cash. Don't. Get a separate carrying case for your credit cards and put them in an inside pocket that is well-secured and deep. Somewhere, keep a record of your credit-card numbers and the name of the issuing credit-card organization. Credit cards can now be protected against unauthorized charges if lost. One phone call to the security agency that has registered your card numbers can stop charges against all your cards. Notify them immediately if your credit cards are stolen. Never, under any circumstances, lend your credit cards to anyone else, and that includes relatives. Do not allow credit cards out of your possession, except for the few moments you need for paying a bill. Then, be sure to remember to get your credit card back. Quite commonly, a man or woman, most conscientious with cash, will walk away after making a purchase and leave the credit card in the possession of the cashier.

Do not flash your credit cards any more than you would with money. In many ways, a credit card is more valuable than money for it is actually an invitation to unlimited spending. Your credit card has provision for your signature. Sign it. That signature may be of some help in preventing the use of the card in the event it is stolen.

The best credit card is one that has your photograph as an additional form of identification. Unfortunately, many persons receiving credit cards in payment for bills fail to make the necessary check between the person holding the credit card and the picture and, equally often, do not verify the signature.

TAXI SHARING

After leaving an airport, or a train terminal, it is tempting to share a cab, particularly if a taxi is difficult to get, and especially if the weather is rather bad. You may be asked to become a communal cab passenger by someone you just met on the train or plane, or it might be someone at the taxi station. Consider, however, that a taxi is quite unlike a subway, trolley, or bus. The taxi driver and the other passenger may form part of a holdup team. You can easily be separated from your wallet and your luggage and then unceremoniously dumped in a strange part of the city.

There are two basic types of taxis. One is operated by a reputable business organization; the other is known as a *gypsy* cab. The gypsy is an independently owned cab, and may not be registered with the city authorities. And it may simply be cruising in an effort to get some "easy pickings." This doesn't mean that all gypsy cabs are operated by thieves, and that all cab drivers of registered taxis are shiningly honest. The odds, though, are not in favor of the gypsy. If you plan to use a taxi you'll find it helpful to be able to distinguish between the gypsy and other cabs. Spend a few minutes looking at the taxicab traffic and you'll soon get to know which is which. A bus or trolley may take longer, but it is safer. And so is an airport bus.

SLEEPING AND THEFT

During a trip, whether by bus, train or plane, the motion of the vehicle, plus the fatigue produced by traveling, often induces sleep. Again, this is an open invitation to theft. The person sitting next to you, with the help of a garment such as a coat, or a newspaper if the weather is warm, can easily open a pocketbook completely undetected. Further, you will not know of the theft until sometime after it has happened, and while you may suspect that the person sitting adjacent to you was responsible, you will never be sure.

The best place to sit is immediately adjacent to a window, and not an aisle seat. With a window seat you can put your pocketbook between yourself and the window, putting the thief in a position of having to reach across you. This doesn't mean that it can't be done, but at least you have put an obstacle in the way. Do not put the pocketbook into its position by itself. Keep your arm through the straps, and give the pocketbook one twist so that the straps do exert some slight pressure on your arm. Then put a garment, such as a sweater or coat, over the pocketbook so that it just doesn't sit there as a direct temptation. The trouble with an aisle seat is that you must either keep the pocketbook on your lap, or between yourself and an adjacent stranger. Your lap is the better location, but again, give the pocketbook one turn to twist the straps and then cover it with a garment. Naturally, if you are carrying cash or valuables you do not want to lose, avoid opening your pocketbook. If you need chewing gum, lipstick, a mirror, or whatnot, take these items along in a small carryall designed just for that purpose.

If you are being met at a terminal by someone who is a complete stranger, get a description before you make the trip. As a further check, ask a few leading questions that will help make the identification a positive one.

CHECKLIST FOR TRAVELING SECURITY

The unfortunate part about security is that it requires unending vigilance. For maximum protection you must exercise constant alertness and awareness so that you may not become a victim. It is unfortunate since it requires you to be suspicious of people who may very well be honest and interesting companions and who, possibly like yourself, are reaching out for human contact without the slightest thought of possible gain. It also means that instead of concentrating completely on your business or your travel plans, you must devote some time to an activity that, by its very nature, puts a sort of wall around you. The alternative, though, from a viewpoint of financial loss, physical harm, and shock, is a very high price to pay.

Before you go on a trip, consider what precautions you should take before leaving and during the trip.

1. Make a record of your credit-card numbers. Keep a list of them in a small pocket notebook so that you can supply these numbers to the issuing credit organization without delay. Your personal liability, or its extent, may depend on how quickly you do this. And, before you leave, supply a duplicate record of the credit-card numbers to your office secretary or your wife or to someone else who is equally dependable. Thus, if your credit cards are lost or stolen, and if your own personal notebook record disappears at the same time, you can get the information you need just by making a telephone call.

Never let your credit cards out of your possession for any longer than it takes a cashier or sales clerk to make up your purchase slip. And, when doing this, do not let yourself be diverted by conversation, or by thinking about your travel plans. Get your credit card back and make sure it is in a separate wallet, placed in a hard-to-get-at, deep inner pocket.

2. Before you leave, make a realistic estimate of the amount of cash you will need to take with you. If your travels include visiting a particular town or city on a regular basis, consider the advantage of opening an account at a bank in that city, preferably located near your hotel. In this way you will be able to reduce the amount of money you feel you must have with you.

Your currency should have the smallest, convenient denomination. Flashing a hundred-dollar bill for a five-dollar purchase may be impressive and it is, for it will undoubtedly attract attention. The two-wallet idea is a nuisance but it is a helpful security measure. Keep your credit cards and travelers' checks in one and your currency in another. For your currency wallet, select a wallet that permits the currency to be flat at all times, thus producing a less noticeable bulge in your jacket. Your clothing will look better for it. Never put your wallet in a back hip pocket. It is very convenient to keep currency in a pants pocket, held together with a money clip. But this means you can lose it to the first skillful pickpocket who comes along.

3. If you carry a briefcase attach an identification card to the handle with your name, address, and phone number clearly printed. Put a similar identification inside the briefcase. If the briefcase is to contain valuable, confidential papers, make sure it can be locked. Keep it locked, even if it is to be placed next to your seat in a plane or train. The same sort of thinking applies to an attaché case. To avoid the nuisance of keeping and using a key, consider briefcases and attaché cases that use combination locks. These can generally be set to any number you indicate. Select a number that has meaning for you, so you won't forget it. If, for example, your birthday is January 25, then you can make the lock combination 125 (with the digit 1 representing the first month of the year).

Many briefcases and attaché cases look somewhat alike. As a quick aid to identification, have your initials affixed. If you have a choice of initial styles, just make sure they are sufficiently large so they can be read easily. Some initials, made in an Old English font, are quite attractive, but are also almost impossible to read.

Do not carry more confidential business information in your attache or briefcase than you absolutely require. The best place for such papers is in a safe.

4. If you use limousine or bus service and your baggage is to be loaded, generally into a side compartment in the bus, or the rear

of a limousine, make sure that your luggage is loaded. Do not take your bus or limousine seat until it is. If your luggage is equipped with identification stripes in color, it will be easy to recognize. If your luggage remains on the sidewalk when your bus or limousine pulls away, you will be quite fortunate if you see it again.

Do not give in to suggestions from any bus or limousine driver that your luggage will follow on a subsequent car. It may, or it may not. If for some reason your luggage must remain to wait for the next available vehicle, wait with it.

5. If, during traveling, you make the casual acquaintance of someone sitting next to you, do not reveal anything of your traveling plans, or the name of your hotel. The queries you get may be innocent, but then again, they may not be.

6. If, as part of your travel plans, you rent a car, treat it as though it were your own and exercise the same precautions you do at home. Don't park on a dark street or a street that isn't adequately lighted. If possible, park in an inside garage, even if there is street parking space available. Do not leave anything more than the ignition key with the parking attendant.

Obviously, it isn't always possible to follow these suggestions. If you must park on a dark street, don't rush to the car on your return. First, look to see if no one is lurking about. A car thief will sometimes wait for the return of the car owner or operator, and will then not only steal the car, but rob the owner or operator as well.

Again, if your rented car is parked on a dark street, have the door key (this is sometimes the ignition key as well) out and ready to use. You should be able to walk up to the car and open the door immediately, without fumbling around in your pockets. Do not remain at that spot for any longer than absolutely necessary. Drive away at once, and then, if you must examine some papers or stop for any other reason, select a well-lighted area, preferably one with ample street traffic.

The situation is much more difficult for women. A woman is likely to put up less of a defense than a man and so is a more attractive target. Further, as mentioned earlier, women have a habit of waiting until the last possible moment before opening a pocketbook. Remove your car keys from your pocketbook and have your door key ready long before you walk down a dark, deserted street to your car. And drive away as quickly as you can. If possible, try to arrange for an escort to your car.

7. Do not pick up hitchhikers under any circumstances, not even if the hitchhiker is a woman carrying a baby in her arms. This ploy has actually been used. Any male hitchhiker is a potential thief and a potential murderer. Police files are filled with records of kindly drivers who ended up robbed and killed.

Don't stop your car to ask directions of pedestrians. And be suspicious of such an answer as "I'm heading that way myself. Let me go along with you and I'll show you exactly where it is." Instead, drive into the nearest service station. They get requests for directions quite often and are usually familiar with the various roads.

8. Don't start out with a road map and fond hopes that you will arrive at your destination in due course. Instead, plan your trip so that you avoid getting to your destination late at night, particularly if you must drive through or stop in some high-crime area. If for any reason you are menaced while you are in your car, turn on your emergency flashing lights, and keep beeping your horn. Don't lean on the horn so that it emits a steady blast. Most people will think your horn is stuck. A series of horn blasts will attract the kind of attention you want.

9. When you get to your hotel or motel, put all valuables you do not require in the hotel or motel safe. This doesn't mean the safe cannot be robbed. And it doesn't mean that the hotel or motel proprietors assume full responsibility for your valuables. It does mean that your property is more protected in a safe than in a motel or hotel room. Ask for safe privileges when you check in and deposit your property at once. Don't wait until you get to your room. However, if this isn't possible or practical for you, do so as soon as you can after you are in your room. If a bellboy accompanies you to your room, ask him to wait a minute or two to accompany you down to the safe-deposit area.

If you do have valuables, don't try to hide them in your hotel or motel room. First, there aren't that many available hiding places and second, you can be sure that any experienced burglar knows where to look—certainly much better than you do. Never leave cash, credit cards, or blank checks in your hotel or motel room. Do not leave your wallet or jewelry on a dresser or bed where it can be seen, and don't imagine your possessions will be safe just because you store them in your luggage, even if you lock the luggage. Luggage locks will not stop a professional thief. They are easy to open and all they can do is to ensure that your luggage will remain closed in transportation and to discourage the inexperienced opportunistic thief.

10. Your first step in a hotel or motel room should be to inspect the lock on the door and to make sure you understand fully how it works. When you are in your room, don't assume you are safe just because the door is closed. Most rooms are now equipped with deadbolt locks that can be operated from the inside only. Your first action in the room should be to close the deadlock. You will also probably find a lock with a chain permitting you to open the door a few inches, but with the chain preventing forcible entry. When answering the door, make sure this chain lock is in position before you open the door. If you have any doubt about the person on the other side of your door, telephone the desk clerk. Do not open the door for people who claim to be plumbers, television-set repairmen, or package-delivery messengers. Have them wait while you telephone the desk clerk for confirmation. Don't ever let yourself be panicked by someone at your door claiming to have an urgent message for you requiring your immediate attention or signature.

11. If, upon return to your motel or hotel, you see someone acting suspiciously in front of your door, or if there are one or two men who seem to be working on it, return at once to the main floor and report your suspicions to the desk clerk.

Do not use stairways in a hotel unless you go up and down with a group. Use operator-type elevators in preference to self-service types. Try to avoid going into an elevator alone. If necessary, wait for a group.

When you register in a hotel, be sure to ask for a room on a lower floor, such as the second or third. Thus, when getting into an elevator, first making certain it is occupied by other people, you have an assurance they will at least accompany you in the elevator as far as your floor. If, for example your room is on or near the top floor, it is quite likely you will be alone for at least the last few floors. Hotel thieves are aware of this and so take an "up" elevator for the last few floors, hoping to find a victim alone in an elevator.

11. Before going away from your room, try the door to make sure it is locked. Don't assume this is so just because you closed the door. Try the doorknob and twist it to make sure the door is really locked. When leaving your room, don't turn off all the lights. You should have one or two lamps turned on so that the

room is fully illuminated. A thief can walk down a hotel or motel corridor, and just by looking at the bottom of the door can get an idea if the room is occupied. A complete absence of light indicates that the occupants are either out or are sleeping. Before you leave your hotel or motel room turn on the radio or television set to give the impression the room is occupied. If you have a travel lock with you, and you should, make sure it is inserted in the door properly by following the manufacturer's instructions.

If, on your return, you find your door open, do not enter the room. Instead, go to the main desk and notify the desk clerk. Your room may be occupied by a cleaner, but then again it may not. A well-organized hotel or motel knows where its cleaners are at all times. And don't feel embarrassed if the door to your room is open for some perfectly legitimate reason. It is much better to be embarrassed than to be robbed.

12. Never invite a stranger to your room. Having more than one stranger visit your room can be a disaster.

13. If you are traveling by car always park as close to your hotel as possible.

14. Do not leave articles of value such as pictures, cameras, clothing, etc., in plain view in your car if you must leave the car for any great amount of time, especially overnight. Put these in the trunk of the car or take them with you.

15. If possible, when traveling by car, avoid rest areas at night, unless the rest areas seem to be well occupied by a number of cars. Never plan to stay overnight at a rest area along the road. In some states, it is against the law to do so.

16. If you are in a town or city that is new to you either check with friends or with the hotel or motel management about safe areas and areas to avoid. In some sections, it may not be wise to walk on the streets after dark, or during the early evening.

Even if you receive assurances that you will be absolutely safe on the streets at all times, stay out of alleys, no matter how quaint or historic they may seem to be. Don't take a shortcut, if that shortcut means you must walk across a vacant lot that isn't lighted. When walking down a street, stay on the pavement alongside the curb to avoid being pulled into a doorway or hall.

If, when walking down a street, you see an individual or a group approaching on your side and you sense the possibility of trouble, walk to the other side of the street. If you must walk down a dimly lit street, use the roadway normally reserved for cars and try to walk in a direction opposite that of traffic. In this way you will be able to see oncoming cars.

17. If you walk past a parked car and someone calls to you from it, ignore the call and any remarks that may be made. And if, while you are walking, someone stops in front of you and asks for a match or wants to know the time, don't stop. Just keep on moving. If you hear any strange noises coming from a car, from a side street, from inside a house or building, pay no attention. These are all devices used for stopping pedestrians and maneuvering them into a position or area where they can be robbed, assaulted, or both.

18. If you are held up, consider that your life is worth far more than your money or jewelry. The average person is no match for a man with a knife or gun, particularly if the holder of the knife or gun is high on drugs, has a resentment against society, or is sadistically inclined. An experienced person might be able to appraise the thief quite well, but the majority of persons cannot. Thieves, particularly those who specialize in hotel rooms, usually work in pairs. If you are held up by a pair on the street or in your hotel room, you are not only outweaponed but outnumbered.

19. Some hotels and motels are now posting warning signs, alerting their customers against specific types of theft that have become common. Read these signs and follow their instructions carefully. They are based on experience.

This does not mean that if you follow all these suggestions you will be absolutely safe on a trip. It does mean that you get more favorable odds. There are very few deterrents you can use when on a trip. At home you may have an alarm system or a dog, or you may have heavy deadbolt locks. Also, at home you know which are the relatively safe areas and you recognize which are the areas to avoid. You don't have this advantage in a strange city. You are also at a disadvantage if the section in which you live has a very low crime rate. This may make you careless about personal security and it may make you somewhat more trusting of people than you should be.

Being robbed while you are on a trip means more than the loss of money or possessions. The shock may make you incapable of conducting your business. If you are on a pleasure trip, your intitial reaction will be to cancel the trip and get home as quickly as possible. It will also make you hesitate about taking other such trips. And so a robbery can affect you in many ways.

Fig. 7-2.

Ready Rescue unit, CB radio. (Courtesy Midland International Corp.)

We cannot emphasize often enough that one of the biggest problems for the average individual is acquiring security-consciousness, and that too many of us have an "it can't happen to me" attitude. However, with an increasing crime rate, the chances are that it will happen to you. Why become a statistic?

20. At one time a wallet, whether used by a man or woman, was just a convenient place for storing money in the form of bills. But the usual wallet today contains one or more credit cards (usually more), a driver's license, car registration, telephone credit card, membership cards in various organizations, possibly a library card, various IDs, etc. The wallet is literally a filing cabinet of a large segment of your personal life.

It would be literally impossible for anyone to remember all the cards and their identification numbers. This puts you in a difficult situation in the event the wallet is stolen.

A way out of this mess would be for you to make a record of the contents of your wallet, excluding a listing of the amount of cash you have in it since this is such a variable, and to keep that list in a safe place in your home, or in your safety deposit box.

CB RADIO

A Citizens Band (CB) transceiver, a combination transmitter and receiver, is a useful security device for you can use it to call for help if your car becomes disabled along a seldom traveled road. No operating license is required.

One unit, shown in Fig. 7-2, has a built-in antenna adaptor which allows the antenna to be mounted directly on the product for short-range communications. This feature is helpful in situations where leaving the vehicle could be dangerous or inconvenient.

The helical antenna can also be attached to a magnetic mount base and placed on top of your car for longer range use.

Chapter 8
Security in Business

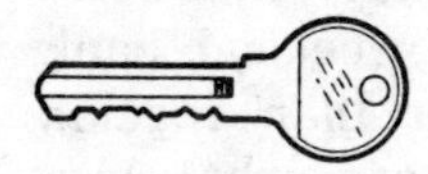

Like houses and apartments, cars, trailers and boats, the business office is a prime target for the burglar (Fig. 8-1). In a sense, though, the office is more like a car from the burglar's point of view. A thief can look inside a car or just glance at it and

Fig. 8-1.
Business office is prime target for the burglar.

he can easily see if it is occupied or not. After a certain hour, making allowance for the presence of members of a cleaning service, offices are unoccupied. Nor are they usually occupied over a weekend, particularly on Sunday, and it is fairly certain that they will be deserted during a holiday.

This doesn't mean that office thefts do not take place during normal business hours. Some thieves prefer the hours of nine to five since they can then use normal building pedestrian traffic as a cover.

At one time, the prime target in an office was the office safe. But with the increased use of nighttime and weekend bank depositories that permit deposits during other than regular banking hours, the office safe has become somewhat less common. More usual targets are now office typewriters, electronic calculators, adding machines, portable duplicators, computers, dictating equipment, security cameras, electronic cash registers, audio-visual equipment, word processors, film and slide projectors, radios, and television sets (Fig. 8-2).

Fig. 8-2.

Office machines are lucrative haul for burglars.

There are some offices, though, that are of special interest to the burglar, and these include those involved in precious stones, currency exchange, silver, and other precious commodities. These offices are ordinarily more secure than the average, but are also a more tempting target.

BUILDING SECURITY

Many building security arrangements are pathetic. Those who man the building after office hours usually double as elevator men, superintendents, or overall caretakers. They seldom receive security training and in many cases the general attitude of the building owner is that office security is the responsibility of the individual tenant. Quite often the owners of the building use a building management service with both building service and owners primarily interested in getting an adequate return on their investment.

In many offices, visitors after hours are required to sign in and sign out. This very crude security measure is often ignored by both tenants and watchmen and, in any event, those in charge of the registration book do not know every tenant. A burglar, dressed for the part—that is, wearing a business suit and carrying a briefcase or dispatch case—and making an impressive appearance can walk with confidence to the entrance desk, sign in, and go to any floor he wishes.

In many buildings, no packages may be taken out without a pass. Passes can be made in advance by the burglar, but in many cases, the recipient of the pass, usually the head elevator operator, doesn't even both to look at it. Since this system obviously offers such little protection, some building operators provide their own printed passes. Theoretically, only tenants are supplied with such passes, but any thief can manage to get possession of one. He can then have the pass duplicated by an offset process by any small printer. In a building having a large number of elevators and a number of different exits, such as from the front and rear, or front and side, the pass sytem falls apart completely. The elevator starters immediately assume that a starter on the other side of the building has asked for and received the pass.

In some buildings packages may be removed only via the freight elevator. The pass system, if in effect here, provides as little security as passes used in regular passenger elevators.

This means, then, that as far as the average office is concerned there is no perimeter protection. The burden of security falls on the shoulder of the business office owner or his general manager.

Protecting the Office

Offices in some of the older buildings have opaque glass doors, a style that was common at one time. With a suction cup and a glass cutter, and a reasonable assurance that he is not going to be seen or disturbed, the burglar can easily cut away a section that will let him reach in and release the door lock. Most office locks are not the double-cylinder type and a key is required only for entrance. Once he has his hands inside the door, the lock is defeated and the door opens easily. The best arrangement with a glass door is to have it replaced with a metal type, with the consent and approval of the building owners or operators.

While the single cylinder, deadbolt lock used for an office is more or less pick resistant, a burglar can defeat this lock with a cylinder puller. For better protection, use a cylinder-guard plate. This will not interfere with your use of your office key and will provide greater security by making the cylinder puller ineffective.

Glass doors can be, and often are, protected by a sensing-foil setup, designed to set off some kind of alarm. Sensor switches can also be used to trigger an alarm if the door is opened. The alarm can be an on-premises type, or the type that signals a warning to some externally located security agency. It is always advisable to have a sticker on the door emphasizing that the door is alarm-protected, whether or not this is actually so.

Three Steps to Office Security

There are three basic steps to office security: perimeter security, office security, and spot security. With perimeter security, the objective is to keep the burglar away from your front door. Perimeter security is a function of the building owners' operators and there is nothing much you can do about this, unless you make inquiries before you sign a building lease. Perimeter security could include adequate lighting not only in the front of the building, but of the sides and rear as well. Spotlights and floodlights are useful deterrents. Adequate perimeter security also includes a trained guard. For reasons of economy, many buildings use elderly retired persons, often retired police officers, as building guards and watchmen.

Office security means not making your office especially attractive to a burglar. In office security, it is best to keep the burglar away from the building altogether, and this is a function of perimeter security. Failing that, the next step is to discourage the burglar from entering your particular premises. You can do this by using not one, but two deadbolt-type locks having double cylinders, with a guard plate on each. You can also use an alarm system on your door. If your building is an older type and the door has a transom, nail it shut and replace the glass with solid, thick wood or metal. Make sure your door fits snugly against the door jamb to minimize the possibility of the use of a jimmy. If your door lock is equipped with a lock alarm—and this would be helpful—also make sure that there is a decal right above it stating that the lock is so equipped.

The basic idea in perimeter and office security is to discourage the burglar from entering, by throwing obstacles in his way, and by encouraging him to think that the risk of detection is high. However, the trouble with an office is that anyone—and that means anyone—can enter during daytime working hours. A burglar, visiting an office for some contrived reason, can easily appraise the entrance and determine at a glance the possibilities of a break-in.

The Office Key

The office key is one of the weak links in the office-security system. The office door can be opened by building cleaners, and since this service is handled by building management, there is no way to ensure that the key will not fall into the wrong hands. Further, it is common office practice to have a number of duplicates made of the office key. The president of the business must have one, and so must the office manager. Quite often, the lowest member in the office hierarchy, the shipping clerk or the office boy, is given a key for the simple reason that he is the first to arrive and is expected to open lights, sort and deliver mail, and start the day's activities. The greater the availability of office keys and the greater their distribution, the larger the risk.

Once inside the office, the burglar can usually make away with just about anything he can carry. It is helpful to bolt all office equipment to desks—and this is done in some offices—but it can be a nuisance, particularly if the equipment is to be moved around. Some offices are now equipped with radio receivers, television sets, and may have a bar, in addition to the usual collection of general office machines. The burglar's problem is that some of the loot is bulky and heavy. Further, he has the problem of getting it out of the building undetected. And so, in many instances the theft will consist of a single item, such as a typewriter. The burglar can put this under his arm, walk down the stairs, and take the nearest building exit.

Inventory Record

To help the police in recovering equipment, keep an inventory record of all of it, identifying each component by manufacturer, date of purchase, function, and model number. Also make a note of the serial number. You can limit the disposability of your office equipment by engraving each piece with your company name, using an electric pencil. Some offices use self-stick printed decals, but these can be removed or covered too easily.

Unless desks contain important, private papers, they should not be locked. In forcing desk drawers a thief may do more damage than the possible loss of some paper clips and rubber bands.

TELEPHONE LARCENY

Since the office thief, working at night or over the weekend, may not be in a great hurry, he might be inclined to use your office phone to make some long-distance calls. You can prevent this by using a phone lock, but these are suitable only for dial-type and not pushbutton tone phones.

A key operated lock, as shown in Fig. 8-3, prevents unauthorized use of the telephone. The lock does not hamper the communications capability of the phone since incoming calls can be received, but outgoing calls cannot be made. The lock is easily installed by setting the opened lock into the No. 1 hole of the dial and turning the key. A cam at the end of the lock body secures the unit to the dial preventing rotation of the dial for making a call.

TYPEWRITER PROTECTION

You can protect typewriters by locking them to the desk platforms on which they rest. As shown in Fig. 8-4, the lock secures the typewriter firmly. It cannot be lifted or moved but it can swivel and adjust to the typist's comfort. The free spinning, wrench proof locking cylinder is reinforced and protected by a hardened-steel pry-proof collar. It is also saw proof and pull proof. The lock is released by a 7-pin tumbler key.

Another type of lock is equipped with a locking bar as shown in Fig. 8-5. The unit provides 22 inches of solid steel protection, a combination that provides the effectiveness of two locks. The bar is adjustable to fit all typewriters.

Fig. 8-3.

Keyed lock for telephone prevents unauthorized calls on dial-type phones. (Courtesy Taylor Lock Co.)

INSURANCE

While insurance will never cover your office-equipment losses completely, it does spread the financial burden. Since insurance companies are just as interested as you are in minimizing theft, they can often supply useful information on how best to protect your office. If they have a security division, contact them and take advantage of their experience and advice. If you rent office equipment, make sure you know the extent of your liability for the equipment. At the same time, find out if your office machines are covered by insurance by the supplier, or if that is your responsibility. Similarly, if you buy office equipment on a time-purchase plan, make sure you know just what your financial responsibilities are. Ordinarily, in buying or renting office equipment the entire emphasis is on what the equipment can do. The subject of security may not even be brought up during such discussions.

Fig. 8-4.

Security lock for a typewriter. (Courtesy Bolen Industries, Inc.)

PERSONNEL

As in many department stores, theft can come from the inside as well as outside. Personnel departments screen prospective employees from a function viewpoint—that is, how well the prospective worker can do his or her job. For security, a check on background, character, and possible law violations would be helpful.

OFFICE CHECKLIST

The office manager is usually responsible for office security

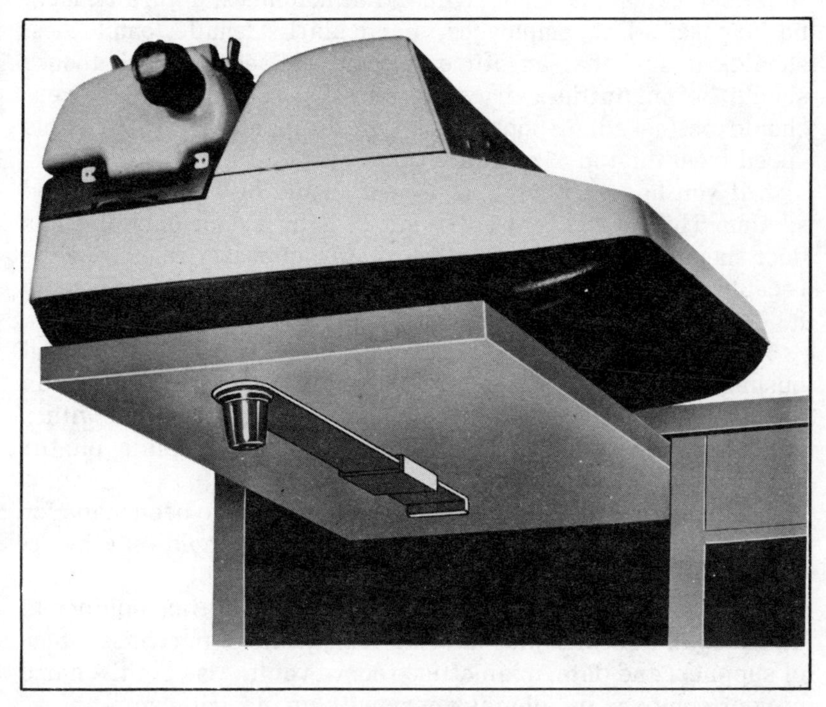

Fig. 8-5.

Locking bar equivalent of two locks. (Courtesy Bolen Industries, Inc.)

but total office security means every member of the staff must be made aware that not only the employing company, but their own personal possessions (including money) can be the target of the office thief. Here is a simple checklist for making the office more secure:

1. Ask questions of any strangers in your area. If you see someone who isn't familiar, lift the phone and alert the office manager or the floor supervisor, or your immediate superior. One of the characteristics of the office thief is his disguise: He looks and acts like a business executive, and he acts bold and sure.

2. Do not allow soliciting of any kind. Most office solicitors are impervious to any rebuffs. The best way to handle them is to phone the building supervisor immediately and complain. Many office-building managements post signs forbidding the use of their premises by solicitors.

3. Keep valuables out of sight. Office machines, naturally, must be exposed. But employees, particularly female employees, should not feel that an office is positively secure. Pocketbooks should be put in desk drawers, and if possible those drawers should be locked. A pocketbook can disappear with incredible speed from the top of a desk.

4. If you have a reception area it should be kept as a closed section. This means that entry into it is by a door only and this door should be kept locked with some sort of buzzer release by the receptionist. The reception room should always have someone in it.

5. Always be sure to lock all entrance doors at the close of business. Some businesses rely on building security and use an ordinary-type latch lock on their exit door. Replace this with a double-cylinder deadbolt lock having a security plate on the outside.

6. Keep a record of the number of keys that have been made for the main office door and make sure these are held only by responsible individuals in the company.

7. Make a record of all serial numbers of all office equipment and keep this record, plus information on date of purchase, name of supplier, and date, in an office-record vault. Also put the name of your company on all machinery with an electric pencil.

8. Warn all employees never to keep personal cash in their desk drawers.

9. The extent of your security efforts depends directly on what you want to make secure. If your office does not handle cash, has no valuable papers or equipment, and has nothing of any value for a thief, and pays its employees by individual checks, then security can be minimal. If you have a safe for bookkeeping records only, put a note on the outside of the safe making such a statement.

10. Consider the possibility of having your office wired with a silent alarm to some external security agency.

BUSINESS SECURITY DEVICES

Various devices can be used for business security including antibugging components or devices that detect the use of bugs, electronic countermeasures, communications security equipment, shielded conference rooms, special alarms, telephone or data scramblers, voice stress analyzers, low light video, bomb detectors, bomb disposal, vehicle security, audio and video-recording products, closed-circuit television, two-way mirrors, etc. An extraordinary amount of countermeasures is available, developed by security professionals with decades of training and experience in the field of counterintelligence. While security equipment of high sophistication is used in business, it also finds applications by governments, national defense systems, medical and research professionals, as well as security agencies and private individuals.

Much of the information on the design of high level security equipment is privileged and understandably the manufacturers of such equipment do not release it, even to their customers.

CLOSED-CIRCUIT TELEVISION

Closed-circuit television, abbreviated as cctv, is easy to install and easy to operate. A cctv system can consist of one or more cameras, either fixed in position upon a particular area such as a doorway, or swiveling to cover an area, as in a group of tellers' positions in a bank. The camera is connected via cable to a monitor having a television screen, commonly about 9 inches in diameter. The system can include an interphone and a camera stand. A cctv system is suitable in private homes, factories, warehouses, supermarkets, stores, or any area where a surveillance or a communications system is required (Fig. 8-6).

The scene in front of the camera lens is converted into an equivalent video signal which is sent to the monitor where it is viewed on the screen. A monitor is like a television set except it is much simpler. It does not have a front end tuner since there is no need to pick up broadcast television signals. The monitor contains a video amplifier, comparable to the audio amplifier in an fm/am broadcast receiver.

The video amplifier strengthens the signal picked up by the camera and delivers that signal to the picture tube. Quite often the cctv system is monochrome only since its purpose is not entertainment but identification. Elimination of color lowers the cost of the camera and simplifies the circuitry in the monitor.

In a representative cctv system (Fig. 8-7), the camera has an automatic light-compensation circuit that ranges from 20 luxes to 50,000 luxes and can be adapted to a wide range of light conditions. The camera can be operated under practically any light condition without additional equipment.

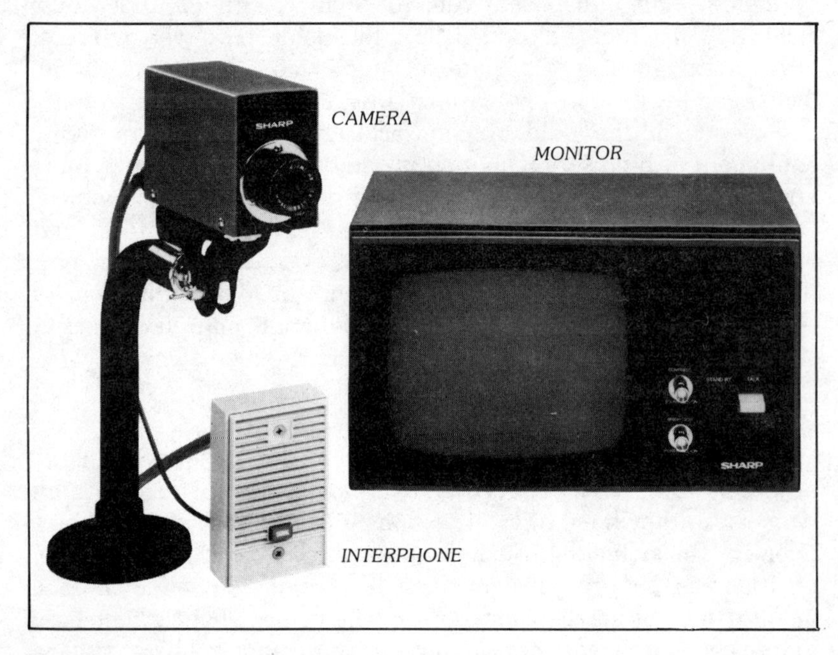

Fig. 8-6.

Elements of a closed-circuit television system. (Courtesy Sharp Electronics Corp.)

The camera will work satisfactorily within a temperature range of minus 4°F to 113°F (minus 20°C to 45°C) which covers practically all normal applications. The system has to be switched to on or standby when used at temperatures below 32°F (0°C).

The camera is equipped with a camera cable having a length of 33 feet (10 meters) through which all voltage and signals are fed between the camera and the monitor. Power for the camera is supplied from the monitor through the cable, hence no connection to a power outlet is required for the camera head.

The cable length can be extended to a maximum of 164 feet (50 meters) while maintaining good picture quality. The system can be modified to form a two or three camera selector unit with selector buttons enabling the user to choose the picture required.

Audio power output is supplied by an Interphone which can also be used as a two-way voice-communications system. The Interphone is equipped with a press-to-talk button. Controls on the front of the monitor include a contrast/sound switch, a bright-

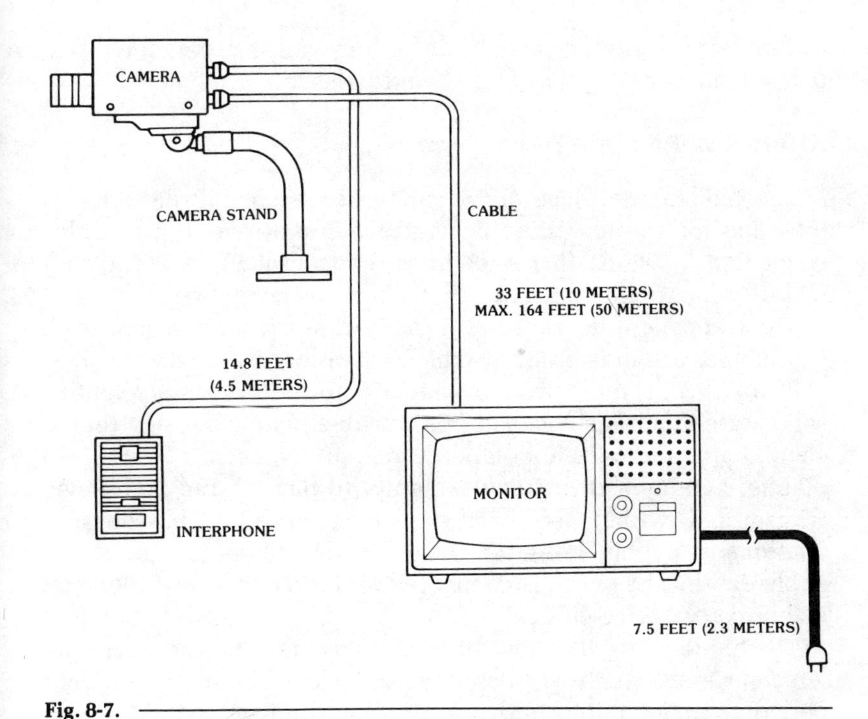

Fig. 8-7.

Closed-circuit television system. (Courtesy Sharp Electronics Corp.)

ness/power on-off switch. A control at the rear of the Interphone is used to adjust sound volume and camera electric focus.

ACOUSTIC NOISE GENERATOR

An acoustic noise generator produces broadband audio noise on an object, typically a window or wall that may be subjected to eavesdropping by spike microphones, hidden microphones or microphones equipped with extender tubes. The unit (Security Research International) is also effective for the disruption of eavesdropping that is in the form of microwave or laser reflection. Since these techniques require that the laser or microwave signal be reflected or bounced from a fairly stable surface, usually a window, the generator defeats these techniques by producing unfilterable noise that masks the entire human voice range by saturating the surface with various vibrations.

Various transducer adapters are included with the unit for coupling to walls, ceilings, air ducts, plumbing, and windows. For a typical installation, such as a drywall or paneling, one transducer per 6 square feet should be used.

HIDDEN WIRE LOCATOR

The hidden wire locator (Security Research International) is intended for the identification and tracing of wires and cables concealed in walls, floors, ceilings, underground, etc. The unit will also locate line pairs at a telephone terminal strip.

The component, a transmitter, is solid state, battery powered. It produces a signal, either warble or steady tone, for the receiver to track throughout a wiring system. The receiver is a sensitive solid-state amplifier encased in a durable plastic housing that is small, light, and easy to handle. A tone probe, encased in molded plastic, acts as a directional antenna to pick up the tone being transmitted within the wiring system. In addition, the unit contains a bubble level for accurate determination of buried cable depth. The probe is connected to the receiver by a flexible, insulated two-wire cable.

The hidden wire locator can be used on many multiple types of wiring systems such as phone cables, television, intercom, door bell, thermostat, public address, etc. Operation of the hidden wire locator is simple. The transmitter clip leads are connected to the wires to be traced, and the receiver is used to scan along the wall or surface concealing the wire, thus enabling the operator to locate the line/signal throughout its entire path.

TAPE RECORDER AND RF BUG DETECTOR

The abbreviation for radio frequency is rf and a bug is any device that is used to pick up voice or other signals. The radio frequency transmitter and tape recorder detector is used to determine if a tape recorder is being used or if an rf bug is present and, as such, is a countermeasure device. The unit, shown in Fig. 8-8, can also be built into office equipment such as a lamp or a barometer to keep its presence secret. The unit is battery operated, is equipped with a high sensitivity room antenna, and has a self-contained rechargeable battery pack.

There is also a pocket sized transmitter detector available that gently vibrates, alerting you to the presence of hidden transmitters. The unit is advantageous if you do not want the warning

Fig. 8-8.

Radio frequency transmitter and tape recorder detector. (Courtesy Security Research International)

signal to be observed by others, as would be the case of detectors using flashing lights and buzzers. The detection range is nominally 20 feet, with all forms of modulation being detected. You can also get a detector equipped with a LED (light-emitting diode) that supplies the warning.

Using microelectronic techniques, bug alerts can be made extremely compact. Some are designed to fit inside a watch case or a ballpoint pen. Not only is the small size convenient, but it helps preserve the identity of the bug. One way to defeat the purpose of the bug is to have it generally known that a bug is being used.

Protecting yourself against one or more hidden tape recorders is one thing and quite another if you plan to make use of one yourself. If, for example, someone telephones you and records the conversation, he is required by federal law to inform you he is doing so. That same law is imposed on you if you plan to record your telephone talks.

TELEPHONE SECURITY

In today's world, eavesdropping is more common than you think. So, when you discuss business matters on your telephone, it may be essential for you to know if your conversation is private.

You can get a telephone security device (Fig. 8-9) that will alert you to line activated devices, taps, wireless transmitters, extension pickups, and voltage changes. To use the component,

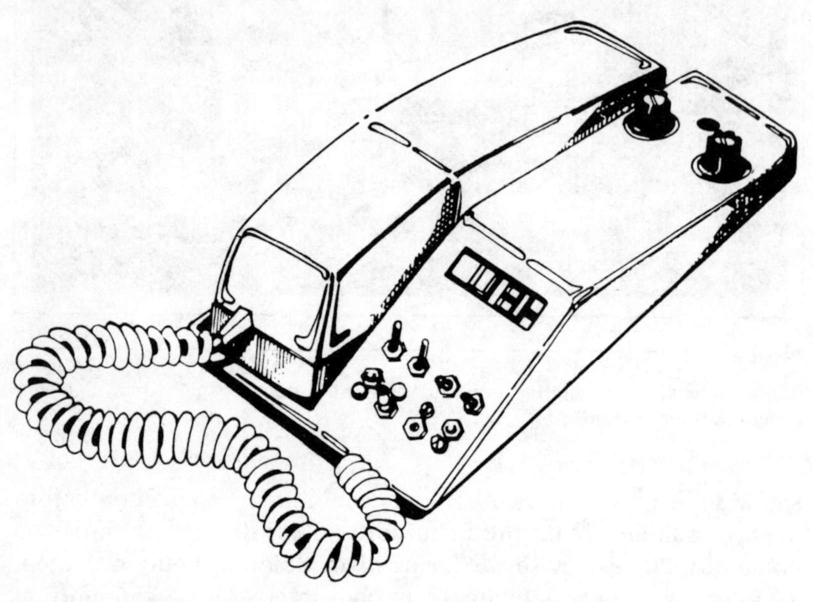

Fig. 8-9.

Telephone security unit will alert you to line activated devices, extension pickup, and wireless transmitters. (Courtesy Security Research International)

simply plug it into your existing wall jack. It will silently alert you to the presence of a direct wire tap, tape-recorder activator and/or switching circuit, telephone tap transmitters (whether parasitic or self-powered), infinity transmitters, extension off-hook alert, and rf detection of any transmission device that may have been carried or secreted in your area. The unit will also instantly deactivate any device on the line, allowing you then to continue your conversation in privacy. A continuous LCD (liquid crystal diode) line-voltage readout allows continual visual monitoring thereby alerting you if significant increases or decreases occur.

PORTABLE TELEPHONE SCRAMBLER

Corporate spying and the stealing of ideas costs business millions of dollars annually. The weakest link to the safeguarding of private, confidential, or secret information may be the telephone.

The portable telephone scrambler is a multiple coded, self-contained, readily portable instrument designed to permit private telephone communications between a pair or more of users. The instrument converts normal speech into unintelligible gibberish by interchanging the high and low tones before transmission over the telephone circuit. A similar receiving instrument again inverts the speech tones thereby converting the gibberish appearing on a telephone circuit into clear speech for the desired listener.

The exact translation of the speech tones is controlled by a tone generator within the scrambler instrument making possible a number of different scrambling codes by varying the pitch of the translation tone generator. One portable telephone scrambler provides individual speaking and receiving selection of 25 different translation codes on the control panel of the instrument. The letter-coded switch controls the speaking translational tone while the numerically coded switch controls the receiving tone translation. The equipment can also operate, through repeater stations in a base to base mode, base to mobile or mobile to mobile.

KEYED SHOWCASE LOCK

A keyed showcase lock, as shown in Fig. 8-10, provides protection for valuable items in glass showcases. The lock secures two sliding panels at the center point where they overlap, preventing either from opening. A locking bar with a U-end is positioned over the edge of the inner panel and secured by tightening a screw. The serrated portion of a locking bar lies flat against the outer surface of the inside panel. The lock is then slipped on the locking bar against the edge of the outer panel, locking it into place. Use of the key is required to slide the lock off the bar.

ALARM SYSTEMS

Alarm systems, previously described in Chapter 4 for house and apartment use, are also available for business protection.

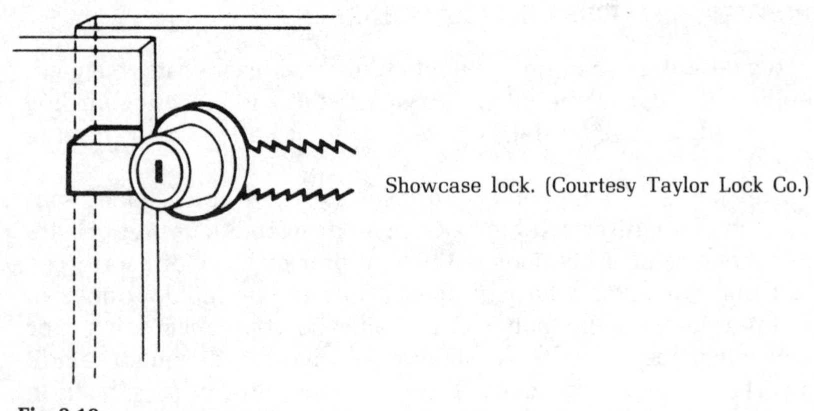

Showcase lock. (Courtesy Taylor Lock Co.)

Fig. 8-10.

However, flashing lights and a warble-type alarm would be useless if a break-in took place during a holiday or over a weekend, particularly in a warehouse or a relatively unoccupied office building.

For business, a possible arrangement would be one or more perimeter-type devices connected to a 24-hour security office.

WINDOW FOIL

Window foil is commonly used to protect store windows, but it can be used for any large area of glass, such as an inside showcase. As shown in Fig. 8-11, this is simply a strip of narrow

Fig. 8-11.
Foil is commonly used to protect store windows.

metallic foil cemented around a glass window and forming a continuous conductive path that can become part of a series electrical circuit. The foil works as a sensor switch. If the window glass is shattered by a burglar trying to enter that way, the foil will be broken, opening the circuit and sounding the alarm.

Window foil is also supplied with a self-stick backing, making it easier to apply the foil. Foil can be used to protect the rear windows of a store, possibly facing an alley or some other area with little or no traffic. Fig. 8-12 shows how the foil is used this way for an upper and lower window. The inset in the drawing shows the method of splicing the foil. The other inset indicates the correct method for making a foil corner. The heavy black lines, terminated in an arrow, represent the wires leading away from the sensors to the alarm system.

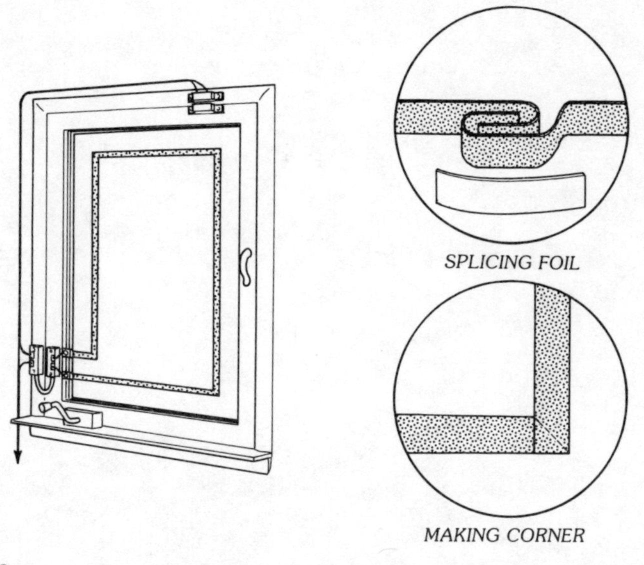

Fig. 8-12.

Window-foil installation using magnetic switch sensors for casement windows. (Courtesy Hydrometals, Inc.)

Chapter 9

Bunco

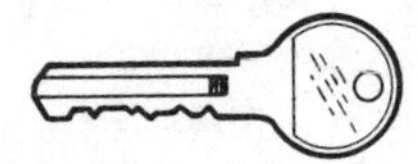

Every crime requires a minimum of two persons: the perpetrator and the victim. Even if the victim is not present at the crime, it is his cooperation that makes the crime possible. The smaller the extent of cooperation, the less likely it is that the crime will reach a successful conclusion. An unlocked car door, an apartment with inadequate locks, careless walking at night through heavy crime areas—these are all examples of cooperation that invite crime.

There are two types of cooperation, active and passive. In passive cooperation, the thief tries to avoid confrontation with the victim. Someone who steals the tires off your car is probably appreciative that you parked your car in a dark area next to a vacant lot, but he would be less appreciative if you made an appearance while he was at work.

A crime in which there is active cooperation is called by various names, bunco, scam, or confidence among them. Many crimes of this type are based on a well-known but highly prevalent human failing, the desire to make money without working for it. There is a saying that we all have some larceny in us. It is this larcenous streak that makes us susceptible to bunco-related crimes. Oddly enough many of these crimes are designated as "games," such as the Bank Game, Lemon Game, Easy Money Game, etc.

Bunco schemes are so common, so varied and so extensively used that the police in some of the larger cities have trained personnel in a group called the Bunco Squad.

PIGEON DROP

There is no racism in crime, nor is crime restricted by virtue of creed, nationality, age, or sex. Criminals can be of any age, any color, male or female, be citizens or aliens, and can have any religion or no religion.

Pigeon drop is generally operated by females, either white or black, or operating interracially. The victims of this bunco game are usually elderly women and can be women of any race.

Although this bunco has many variations, it generally adheres to the following guidelines: The victim is selected in a bank or an above-average neighborhood. The first bunco operator, appearing quite sociable, engages the victim in conversation (Fig. 9-1). At this time the second bunco operator appears on the scene and finds an envelope on the floor. The first operator calls the attention of the victim to the find.

Fig. 9-1.
Pigeon drop.

The finder of the envelope, the second operator, opens it and is startled to find that the envelope apparently contains a large sum of money. Since this game is played out in the direct presence of the victim and her newly found friend, there is no way in which the envelope's finder can disguise or hide her newly acquired riches. Consequently, she approaches the victim and the first bunco operator, and says: "You saw me find this envelope, didn't you? It doesn't have any name on it. Look! It must have eight or nine thousand dollars in it. What in the world are *we* going to do with it?"

Note the word *we*. It is with the use of this single word that the bunco operator ties all three of them together. And now the victim begins to rationalize mentally; she was present at the find, her presence was acknowledged, and "finders are keepers" becomes an acceptable philosophy.

Now the two operators start a discussion about various ways of handling the money. At this time, one of the bunco operators suggests that they contact her employer, an attorney. She is certain there would be no legal fee, for she is a long-time trusted employee. The three women then proceed to an impressive looking office building where the "employee" takes an elevator to an upper floor, leaving the two other women waiting for her in the lobby. Shortly thereafter she returns and states that her employer counted the contents of the envelope and found that the amount was twelve thousand dollars, and that the three women were legally entitled to it in the absence of any known or rightful owner. However, as far-fetched as this might seem, the "attorney" requested that each of the sharers of this new fortune prove their financial responsibility to him, commensurate with the amount of their expected share.

The first operator immediately states that she has four thousand dollars as proof of her financial responsibility, and offers the money to be shown to her fictitious employer.

Attention then shifts to the victim to show proof of her financial responsibility. Reluctantly, the victim produces her bankbook showing a balance in excess of the required amount. However, because bankbooks are easily acquired and altered, the victim is persuaded to go back to the bank to withdraw the required amount. Thereafter, the three go back to the lobby of the office building to wait while the alleged attorney examines each person's offering of financial responsibility.

To ensure the honesty of each of the three women, the twelve thousand dollars has been sealed in the original envelope and entrusted to the care of the victim. With the trap set, the first bunco operator gives her purported four thousand dollars to the second bunco operator for examination. Shortly thereafter, the second operator returns and states that her employer agrees that she should share in the proceeds of the found money.

The moment of truth has arrived. The victim now gives her four thousand dollars to the second bunco operator as evidence of her financial responsibility.

Now you may think the victim is naïve in parting with four thousand dollars, but consider what has just happened. She has just seen the first operator hand the second operator that same amount to demonstrate her financial responsibility. Also, she is being trusted, for she now has in her hands an envelope containing twelve thousand dollars. And so she turns over her money to operator number two, who promptly disappears into an elevator with it, presumably to show her employer.

The two women, the victim and the first bunco operator, remain in the lobby, but as time passes, the bunco operator becomes fidgety and lets her discomfort be seen. Finally, she suggests to the victim that she wait in the lobby, while she (the operator) looks for her friend. She, too, disappears into the elevator.

The victim is now alone in the lobby, but she still has the envelope. Finally, after waiting for a long time, she slits open the envelope, only to find that it contains nothing but rectangular slips of newspaper.

A pigeon drop is sometimes called a switch and, of course, there are many variations. The victim is the pigeon. If the pigeon had insisted on immediately reporting the find to the nearest police station, the two bunco operators, working in collusion, would have found some good reason to disappear and to search for a less honest victim.

In the pigeon drop, the two bunco operators make every effort to determine the amount the victim has on deposit in her bank. If it turns out that she has no funds, or if she has no immediate access to a substantial amount of cash, the operators find some pretext for leaving, taking their loaded envelope with them.

This bunco game is played on women because women, presumably, are less sophisticated in money matters than men. However, this does not mean that men are immune to bunco, for our built-in tendency to larceny applies to both sexes.

If the bunco operators are caught, the value of the property taken determines whether they will be charged with grand larceny or petty theft.

CREEPERS

This type of bunco scheme is characteristically operated by

two prostitutes or a prostitute and an accomplice. The prostitute picks up her "mark" (victim) in a bar or on the street, and accompanies him to a prearranged location where the accomplice is waiting. The location is a dark, suitably equipped room in which to turn a "trick" (engage in an act of prostitution). After the couple disrobes and while they are engaged in the sex act, the accomplice creeps into the room from a place of concealment, and removes valuables from the victim's clothing, which has been placed on some convenient chair (Fig. 9-2).

Fig. 9-2.
Creepers.

When getting dressed, the victim will discover his loss soon enough, but it is difficult for him to point the finger of blame at the prostitute, for she has been in his company all the time. Even if he persists he will seldom be successful, for the lady is on her home grounds and surrounded by her friends, accomplices, and pimps who are ready to come to her aid. Outnumbered, outshouted, and outtalked, the victim soon realizes he is no physical match for such a group. Again, this type of crime is often unreported. And if reported, all that may happen is that the prostitute will be held overnight, will pay a fine in the morning, and will be back in business the same day, or night. The proceeds from the victim may be more than enough to pay all costs.

BADGER GAME

Men seeking sex are vulnerable to all sorts of games, such as the one just described. Still another, known as the badger game, is usually accomplished by a prostitute and a male friend, and follows a definite pattern.

As a start the lady, often quite attractive, allows herself to be picked up by the mark. She further allows him to make friendly advances and at the proper time, seemingly overcome by his maleness, good looks, and wit, permits him to take her to her private apartment. At the peak of the evening's entertainment, the lady's "husband" returns unexpectedly and is outraged. The husband, though, isn't so overcome that he forgets his camera, conveniently loaded, properly focused, and prepared to take repeated shots of the naked couple.

The victim is now ripe for extortion. His name, address, and business are known to the woman, who has been careful to obtain this information prior to the appearance of her husband. If not, or if the mark has insisted on retaining his anonymity, it is no great effort for the husband to reach for the victim's wallet and extract his business card, and possibly some cash as well.

That cash isn't the end of it, for now the victim is at the losing end of an extortion scheme. Depending on their wealth, victims of such schemes will often pay and pay, rather than run the risk of exposure.

There are many variations of the badger game. A seductive young prostitute, for example, may select one of her higher-priced clientele as the mark. She takes her patron to a particular motel that is decorated in the most fashionable manner and noted for its discreetness. After engaging in sex with the prostitute, the businessman returns to his everyday existence.

A day or two later he receives a telephone call from the manager of the motel, who explains that either Mr. Businessman or his "wife" must have left a cigarette burning in their room, as the motel had a serious fire that destroyed considerable, expensive furnishings in Mr. Businessman's suite. The manager further explains that the insurance-company representative requested him to call to see if a lawsuit could be avoided.

Obviously, the businessman might be inclined to pay for the damages. If he checks the premises, he will find evidence of the fire that will substantiate a lawsuit. Because Mr. Businessman's wife might be upset by the circumstances, situations of this type are often settled out of court.

MARRIAGE BUNCO

Marriage bunco is one of the saddest of all the schemes used for separating people from their money and valuable possessions, for it takes advantage of their loneliness and their desire to build a personal relationship. There are many people who are alone, widows who have been accustomed to a long-term male-female relationship, and young women who seem unable to meet the right man. Such persons often advertise in local newspapers for friends, readily engage in correspondence with strangers, are relatively easy pickups, join lonely-hearts clubs, or attend dances for the sole purpose of meeting someone. They are all ready, and easy, prey for the marriage-bunco operator. Although operators of such schemes can be male or female, this particular type of bunco seems to attract a larger number of men. The male operator is characteristically well-dressed, manages to convey an impression of wealth, is an easy and good conversationalist, knows how to put women at their ease, is sympathetic, kind, courteous, gentle, considerate. The victim, on the other hand, has often had little experience with men, and, as is the case with many widows, may have led a relatively sheltered life. Quite often the woman is a person of modest means, generally with just a few thousand dollars set aside for immediate personal or old-age security.

After initial introductions, and careful interrogation and analysis, the con man chooses his mark. He then sets off on a whirlwind courtship, flattering the woman and giving her the undivided attention she yearns for and that she often believes she is entitled to. At some point, (Fig. 9-3) when the bunco operator feels that his prospect has become adequately "softened," he proposes marriage and is nearly always accepted immediately, without hesitation, without investigation. His own description of himself, of his business, and of his life, is accepted without challenge. Quite often the woman wonders at her tremendous good fortune.

Shortly after they become engaged, the victim is told that although she is loved the actual wedding will have to be postponed until the bunco artist's funds are released from escrow in another city. He explains that he has recently sold his business, and that all his assets will be tied up for about six months pending transfer of title to the new owner. The bunco operator then suggests that the marriage can take place immediately, however, if she will lend him a substantial sum. The bunco operator has long since learned of his fiancée's cash resources and is able to indicate just what the required amount will be.

Fig. 9-3.

Marriage bunco.

The woman generally withdraws her entire savings and gives it to the bunco operator without hesitation. After all, their money will all be in the family once the marriage ceremony is performed. A slick con man can often maneuver his victim into a position in which she proposes the loan, while he demurs and appears to resist but finally capitulates to her demands that he take the money. The more reluctant he appears, the more she insists. Finally, he accepts the cash, and is then never seen again.

There are many variations that bunco operators use in marriage bunco. Some are equipped with fancy gilt-edged stock certificates, or deeds to real estate, or bundles of documents, all of which tend to convey an impression of wealth; not cash, but wealth that can be turned into cash at some future date. The basic bunco remains the same, though—the separation of money from the victim under the promise or inducement of marriage. Marriage-bunco operators seldom marry their victims, since most of them don't want to become involved with bigamy laws in the event they are apprehended. One notorious Los Angeles operator, however, married 26 women and successfully separated them from their savings before he was caught.

MONEY MAKING MACHINE

It would seem that only someone who was naïve or stupid would accept or believe in a bunco scheme and, indeed, some of the schemes are crude. But some buncos are extremely sophisticated and involve considerable preparation and planning. Many otherwise intelligent people have purchased stock in gold mines that did not exist, in oil wells that were never drilled, in real estate located completely under water, in apartment houses that would never be constructed. Quite often referred to as investments, or more blatantly as a once-in-a-lifetime chance to get rich, or an opportunity to "be your own boss," they have worked successfully against all types of people. No individual can claim immunity against a bunco scheme. If the bunk appeals to our vanity, desire to be financially independent, wish to stop working or whatever the need or want may be, somewhere there is a bunco artist ready to take advantage. The bunco operator's problem isn't developing a scheme—that he can do. His big problem is making contact with suitable victims.

The moneymaking machine (Fig. 9-4) is another bunco scheme, and while it seems impossible that anyone in his right mind would fall for it, yet it is used regularly and successfully. As a general rule, this form of bunco is worked by a person of foreign extraction against victims of similar heritage. The victim doesn't just happen, but is selected. Once that is done, the con man begins a campaign to impress the victim, including lavish entertainment and an apparently never-ending supply of new bills. Ultimately, the victim notices that the bills never seem to be old or used and finally asks about it. After considerable urging, the con man reveals the source of his money: a moneymaking machine. The victim is skeptical at first but the bunco operator demonstrates the machine by opening two compartments; into one he puts a crisp new twenty-dollar bill, and into the other he has the victim

Fig. 9-4.

Money making machine.

put a piece of blank white bond paper. Cautioning the victim not to stand too near the machine, the con man begins to operate the dials and levers, increasing the humming sound from within the machine. While handling the impressive array of levers and dials, the con man pushes a hidden lever that causes a secret compartment to replace the one containing the bond paper. After an appropriate time, some sort of signal sounds, such as a bell. The con man opens both compartments; from one he removes the original twenty-dollar bill and from the other another twenty-dollar bill, which has apparently just been printed.

The victim is now encouraged to take this new bill to a bank and to verify it for authenticity. This, after all, is the real proof. The victim agrees and, the moment he learns that the bill is genuine, is hooked.

From this point on, there are various methods that can be used to induce the victim to part with his money. One way is to let him buy the machine. The con man can explain that he has had the machine for a long time and is willing to share his good fortune. Another technique is to explain that the machine gradually deteriorates each of the original bills used, and, to make a run of new money, needs a large supply of new bills to speed the money-making process. The victim rushes to his bank, withdraws all his savings in new bills, and gives them to the con man, who then agrees to have a duplicate set ready in a few hours. Of course, when the victim returns, both his recent friend and his machine are gone.

SHORT CHANGE ARTISTS

Most people don't like to do arithmetic, and they especially do not like it if it involves any mental effort. And so, it is easy to get short-changed in any retail establishment where the cashier is interested in making a few extra dollars. The short-change technique is usually practiced on transient trade, and works best when there is a long line of customers waiting at the cash register. Some people do not even bother to count their change, while others may suspect they haven't received the correct change but are too embarrassed to do anything about it, particularly if they are in a strange area.

There also are, of course, short-change artists who are customers. One example is the con man who makes a 25¢ purchase, but almost immediately changes his mind and increases it to 50¢. He then pays with a ten-dollar bill and gets $9.50 in change. The bunco operator then states he made a mistake and did not want so much change. He produces and retains possession of five one-dollar bills and a five-dollar bill and asks for a ten-dollar bill in exchange. The victim gives the con artist a ten-dollar bill, but does not take possession of the con man's money. The latter slaps the five ones, the five-, and the ten-dollar bill together and asks for a twenty-dollar bill in exchange. The victim gives the con man a twenty-dollar bill and gets the miscellaneous twenty dollars in exchange.

In retrospect, the involved underlying theft might seem quite apparent, but consider that the exchanges in money take place quite rapidly, accompanied by a distracting chatter from the con man. The con man knows precisely what he is doing; the victim is confused and quickly loses track of who gave what to whom.

If you are traveling in a foreign country you can expect to be short-changed somewhere along the line. The short-change operator can be a waiter, a taxi driver, a bus driver, the seller of tickets for a movie or other form of entertainment. These small change ripoff people take advantage of the fact that you do not know the language and are not familiar with the currency. Quite often they will give you your change in the smallest possible denomination consisting of coins and paper money. The operator knows you must examine each coin to determine its value, possibly do likewise with the paper money, reach a sum, and then deduct this sum from the amount you originally supplied. The victim is easily confused, particularly if by now a line has formed behind him.

You may also eat in a restaurant in which the price is fixed and although the notice on the window of the restaurant and on the menu is in a foreign language, the phrase "price fixed" is easily enough understood. You also note that the cost of service is included, so you may conclude you know what the total cost will be. Perhaps. Don't be surprised to get a bill that is quite different than that appearing on the menu and that it now also includes a 15% surcharge to take care of services. If you try to complain you will receive a barrage of foreign language, possibly surrounded by one or more belligerent waiters and the manager. You are outnumbered, outtalked, outshouted, and you have no choice but to pay and retreat.

A variation of the short-change scam takes place in nightclubs, both foreign and domestic. You order a bottle of liquor. The waiter brings it, waves it before your eyes and mentions a number which you have trouble hearing and which you do not understand. When you finally get the bill you are shocked to learn the extraordinarily high cost of that one bottle. If you try to protest you are strongly informed that this is what you ordered and that you were informed of the cost.

BUNCO POLICE IMPERSONATIONS

There are many crimes committed by individuals who use police impersonation as part of their MO's *(modus operandi)*. These scams often take the form of a shakedown racket in which victims are subject to phony arrests by bogus officers. The incident giving rise to the phony arrest is usually set up by the phony cops to pave the way for acceptance of cash bail or the way for solicitation of a phony bribe.

Like other criminals, suspects engaged in police impersonation become specialists in their field and play a convincing policeman role. Most victims, when interviewed, remain convinced they were dealing with real policemen. On the West coast, the bribery or bail-posting schemes are called "shakes" by both the perpetrator and the police. At the present time, there are several phony-arrest routine variations:

Abortion Shake

The abortion shake can be set up in the following way. The operation is set up and controlled in the same manner used by actual detectives. A female operator, posing as the typical unwed, pregnant female, obtains an appointment for an illegal abortion. The abortionist, naturally, is arrested just prior to the illegal operation. Thus, the stage is set for extortion or bribery. In continuance of the phony arrest scam, the mark is allowed to make cash bail, without going through elaborate booking procedures.

Bookmaker Shake

Phony-cop arrest schemes can be adapted to any type of criminal situation. Things go smoothest when the victim is caught in the commission of an actual crime, or when they think they have.

In states where bookmaking is a crime, bookmakers are sometimes ready victims for a phony bribe shakedown for purported police protection. Sooner or later, the hoax is discovered, so such shakedowns are generally temporary, and are usually most successful against novice bookmakers.

In areas where certain activities, such as prostitution, abortion, shooting dice, smoking marijuana, or betting parlors are illegal, since the victims are law violators, they are susceptible to shakedown artists because they are not in a position to call for police protection.

TILL TAP

Till taps aren't common, but they do illustrate the ingenuity used by some con artists. Here is the way a till tap works against a liquor store, although any other kind of store can be victimized. One of the con men enters the store and at that moment the clerk receives a phone call asking to speak to Detective ______. The clerk asks his customer if he is Detective ______. The phony detective admits that he is and answers the phone, apparently having a conversation with his local precinct regarding a possible robbery. The detective then advises the clerk that he has just been informed that the store is about to be held up and that a stakeout is necessary. The detective advises the clerk to take the money out of the cash register and to put it in a paper bag and to put the bag in a designated spot at the rear of the store. The detective then asks the clerk to lend him his clerk's apron or store jacket, so that he can pretend to be the clerk. He advises the clerk that for his own safety it would be best if he left for a short time; there might be a shootout. The clerk leaves, and upon returning finds that both the detective and the paper bag with the cash receipts are gone.

Some bunco schemes are legal; others, of course, are against various laws; some, like the till tap, are outright forms of theft. In a variation of the till tap, the two operators enter a store separately. The first makes a small purchase. Both operators go to the cashier together. The operator who made the purchase will pay for it to get the store owner to open the cash register. After the cash register is opened, the operator will point to some article behind the store owner, indicating that he would want this included in his purchase. When the owner reaches for the article, the operator will insist it isn't the right one and point to some other item of merchandise. But while this exchange takes place, the second operator reaches into the cash register, removes the currency, and leaves the store. On seeing the empty cash register the owner often makes a wild dash after the quickly departing thief, and so the second operator has time to clean the register out completely, not omitting all the small change.

There are numerous variations of this basic operation. In one example, the operator making the purchase will drop some small coins on the counter so that they roll off to the back of the counter. He does this after the cash register is open. While the store owner is busily picking up the coins, the other operator removes all cash from the cash register.

The solution is simple for the store owner. Always keep the register shut except when closing a sale. Don't become distracted by anything that takes place elsewhere in the store. The first instinct of the store owner should be to close the register quickly, then investigate.

Any store owner with just himself (or herself) in a store is always at the mercy of a pair of operators or a group, In the case of juveniles, two of them can keep a store owner so distracted and so busy at one end of the store that he will not notice the disappearing merchandise at the other end.

MAGAZINE SOLICITORS

Door-to-door magazine solicitors are often con men, just as unscrupulous as the man who sidles up to you in the street and asks if you are interested in buying a "hot" watch or ring. Magazine salesmen are often women, but there is no sexual restriction.

The magazine bunco may work alone or together with another solicitor. The usual pitch is that he is engaged in a popularity contest that will enable him to win a scholarship for college or a cash award to enable him to go into business. Rarely will the solicitor admit he is selling magazines, even under the most direct questioning, until he has gained admittance to your house. Once inside, the solicitor expertly scrutinizes you and your home for things that can be used as a basis for his sales pitch. If he spots a religious symbol, he will automatically be of your religion or studying for the ministry, and will try to sell you religious magazines. If he spots law books, he will be studying to be a lawyer. In any event, he will try to find some common ground of conversation that he can use for his sales pitch.

Whatever story the bunco operator may give, his sole purpose is to induce you to buy a magazine subscription. Once you are committed to a purchase, you are then encouraged to pay in cash. Naturally, if you don't have the cash, a check is acceptable, but the salesman will try to induce you to leave the "pay to the order of" section blank, stating that the company name will be stamped in later. Once out of your home, the solicitor will write the word "cash" or his name in this portion, and will immediately cash it at a local store where you are known, or at your bank.

Here is how you can protect yourself from these types:

1. Don't permit any outside door-to-door salesman to enter your home under any circumstances.
2. Never buy on impulse. The bunco artist will try to pressure you with magazines at tremendous savings, or merchandise, or labor services. You will be told it is the "last one," or that you are the only one to have been selected for this bargain, or that they will make their profit through your recommendations, or that this is positively the last day of this sale. Just tell the solicitor to leave his card, or to write his name and address, and you will check him out.

3. If you do permit the con man inside your home, do not leave him alone for a moment. Their usual ploy is to wait in the living room, pretend they are thirsty, and ask for a drink of water. In the short time you will be away, things of value will disappear.
4. If you must buy, do not pay in cash, and never give an incompletely filled check.
5. Never accept verbal promises. Take written statements only.
6. Never, never sign anything. If you are pressured to sign, indicate firmly that you must first consult your attorney.
7. If you are seriously interested in getting a bargain, no matter what the bargain may be, contact the publisher, or the manufacturer, or the distributor directly. They are as close to you as your telephone.

RELIGIOUS BUNCO

"Your doorstep to our pulpit!"

It may never occur to you to question a person dressed in clerical garb, and yet he may just be a con man in disguise. Like all con men, the unscrupulous religious solicitor presents a picture in keeping with the impression he wants you to have. His clothing may vary from a Salvation-Army-type hat to a complete clerical outfit. His props usually include religious literature and he may even have a Bible in his hand.

When you answer your doorbell, the con man explains his noble purpose and asks you to make a donation for the purchase of Bibles for distribution to the poor. On some occasions, the con operator will ask for admittance to your home to pray for your salvation (and for your generosity). After this conditioning process, he will ask you to donate whatever your conscience dictates for "God's work."

But take a look at how each dollar collected is divided: Sixty cents will go to the solicitor who met you at your door as his wages; ten cents to a crew manager who transports and supervises a crew of solicitors who will cover your neighborhood like a swarm of locusts; twenty-five cents are earmarked for business overhead expense, which includes rent for the so-called church, generally a converted store with a few benches and perhaps a podium complete with a pulpit, and wages for the persons providing the corporate name. Only the remainder, if there is a remainder, will actually be used for the purchase of Bibles for the poor.

Although this operation is unethical, it is not illegal. Most of these organizations are incorporated as nonprofit religious corporations; they are legally classified as churches and are exempt from taxation. They are permitted to solicit donations for their church.

How can you protect yourself and still not turn away those who are entitled to your consideration? The simplest method is not to give money to questionable organizations at your door under any circumstances. If an organization is legitimate, the solicitor should not mind identifying himself and supplying his local business address. When in doubt, telephone your local enforcement agency and ask about the organization.

LEMON GAME

Some bunco operators are highly expert in some form of gambling or may be poolroom sharks (Fig. 9-5). The card con man

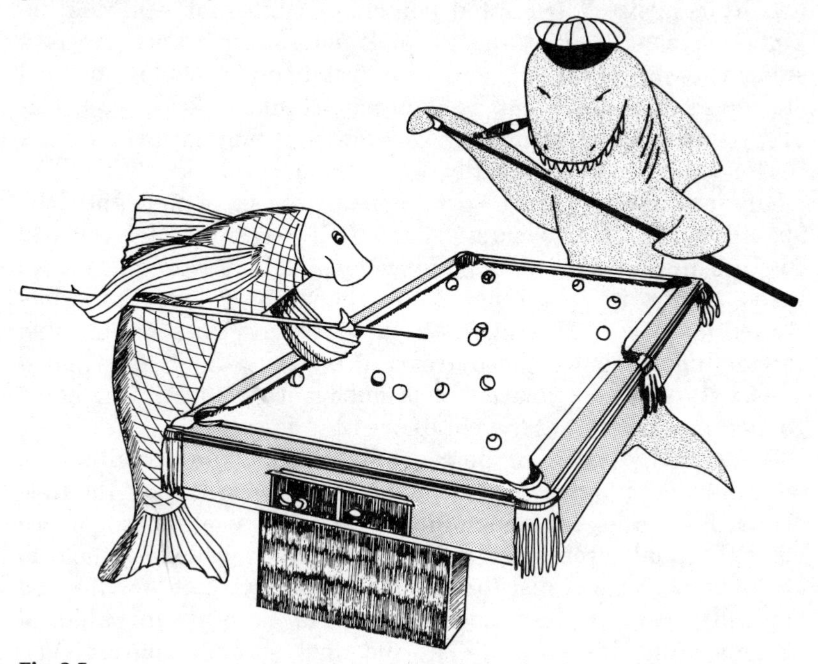

Fig. 9-5.
Lemon game.

may fumble the deck, ask questions that reveal ignorance of the fundamentals of a particular card game, and often show a large roll of bills; he will brag about his ability at cards while making it quite obvious he is a complete amateur. In short, what he is doing, and doing quite skillfully, is putting himself in a position to be taken. This has a double appeal for the mark. First, the mark may be motivated by some slight stirring of larceny at the sight of all the money being displayed. And, second, the con man's attitude is so obnoxious that the potential victim is determined to teach him a lesson. Now, this may be rationalization since greed is probably the main factor pushing the mark into suggesting a game of cards.

Another type of confidence game known as the lemon game involves the use of a pool table. While this type of confidence game isn't too widespread, there are enough amateurs who play a fair game of pool and possibly overestimate their own abilities to make this type of con game possible.

The "lemon man" will have all the main characteristics of a successful bunco operator. He will have a glib tongue and a sharp mind, both essential for someone who lives by his wits. He is usually a highly extroverted person, is congenial, and has the ability to meet people and make friends quite easily. He is a superb actor and not only understands human psychology but is a master at its application. Equally important, he is an expert at playing pool, and is far more competent at it than most amateurs.

The lemon man is usually found in a poolroom, but unlike amateur players, shoots a very restrained game of pool, and does not allow his skill to manifest itself. If anything, viewers will come to the conclusion that he is quite below average, a little better than a raw beginner. More recently the lemon man has shifted his base of operations to the bar that offers a pool table for the amusement of its patrons. The reason for this is that a mark with several drinks in him is much more susceptible to being goaded into accepting the challenge of a game.

In a typical operation, the con man strikes up an acquaintance with the victim and suggests a game of pool, just to pass the time or just for fun, or for some other seemingly harmless reason. To make the game interesting, he then suggests they play for the cost of the game. This is just the opening gambit, though, for the con man proceeds to lose each game. He is quite capable of demonstrating his lack of skill and does so very convincingly. After several games, during which he seems to become more upset after losing successively, he suggests to the victim that they double the bet to give the con man an opportunity to recoup his losses.

Here is where the skill of the lemon man becomes important. He always manages to win, now, by extremely small margins, apparently doing so only by a fluke, while convincing the mark that each game has been won by pure chance. The victim, of course, has long since been convinced he is the superior player and his ego readily regards his growing losses as accidental.

Meanwhile, the bets are increased in size and the victim falls more and more in debt to his "lucky" opponent. By this time, it is now the victim who suggests doubling the bets so he can get even. However, the bunco operator carefully continues to win, allowing the victim to win occasionally to maintain his confidence. If worked properly, the lemon-game operator soon owns the entire contents of his victim's wallet.

In a bunco of this kind, there is very little law authorities can do. When the victim cooperates with the con man, as in the lemon game, there is little the police can do to protect the individual against the consequences of his own actions. The law cannot protect people against their own stupidity.

BANK EXAMINER SHAKE

Elderly women are often the favorite targets of con men and women for several reasons. Such women are generally unsophisticated and have usually been protected for most of their lives, first by a father and then by a husband. Since the male mortality rate is much higher than that of the female, such women are usually widows. In many cases they live on a fixed income such as Social Security or Social Security combined with a small pension, meanwhile keeping their lifetime savings in a bank as a bulwark. They are often lonely and equally often welcome the attention of people.

The phony-bank-examiner game, like the pigeon drop, is generally employed against women of this type. The most frequent sites of these operations are residential areas, especially apartment-house districts, near shopping and small business centers that have local banks.

Confidence men using the phony-bank-examiner scheme work in two- or three-man teams, are usually successful, and often manage to avoid arrest by selecting their victims carefully and proceeding with great caution throughout the entire transaction. They do not hesitate to modify their plan of operation as they go along to assure success and minimize detection. As an example, if the victim contacts anyone other than those she is instructed to meet, the con men will stop their activities and look for a different prospect.

This bunco begins with a telephone call to the victim. The caller identifies himself as a bank official, assuming a role such as an examiner, or an auditor. His first move is to throw the victim off emotional balance. He does this quite easily by telling the victim that the bank has been experiencing some losses due to theft in its accounts, including hers. Completely distraught at the prospect of losing her nest egg. The victim is now given an opportunity not only to recoup her purported losses, but to make a substantial profit as well.

The con man asks the help of the victim in catching the thief and offers a reward of cash or bonds as an inducement. She is also told that the bank examiners are not yet aware that money has disappeared from her account, and that if she makes a substantial withdrawal before they find out, the loss will be the bank's, not hers. If the victim agrees, the con man with some adroit questioning determines the maximum amount the woman has in the bank, and the name of the bank. He then advises her to withdraw a specified amount, not enough to close out the account, but with just a few dollars remaining in it.

The victim is cautioned not to discuss this investigation with anyone, particularly with bank officials, so as not to alarm the thief who has been pilfering her account, and so as to make her eligible for the reward. She is told that if the thief should disappear, she will lose the money in her account as well as the reward. She is also told to get the money in cash and not to handle the money so as not to destroy the suspect's fingerprints.

Originally, the victim was instructed to return to her home, where a detective or bank investigator or official examiner would retrieve the money and return it to her account at the bank. At that time she would also receive her reward for her cooperation. In a subsequent modification of this bunco, the victim is told that just after she leaves the bank an official will approach her and identify himself with a code word or number. The victim would then be required to turn the money over to this person.

There is one unusual aspect to this particular bunco, and it is one reason why the operators are so seldom apprehended. Note that the victim never sees the con men. The only time she does is when she turns the money over to the official meeting her in the street. This meeting is so brief and the victim so excited, not only about assuring the safety of her funds, but also about the substantial reward promised to her, that she is invariably unable to give any description of the person to whom she gives her money.

Again, this bunco seems to violate common sense and ordinary prudence and it would seem that the female victims should know better. Keep in mind, though, that somewhere there is a bunco operator scheming against your funds, and that you, in turn, may be as ready a victim. There is a little larceny in nearly all of us and con men know it. They can tailor any scheme, make it as naïve or as sophisticated as you might imagine, with full knowledge of the psychology and predictable human behavior of their victims.

GOLD SALE BUNCO

Gold has always had a fascinating attraction and for some an almost mystical appeal. Probably one of the earliest of the bunco schemes consisted of painting bricks a gold color and then finding a suitable victim. In some instances the victims were more plentiful than the bricks. It is absolutely amazing what the human mind will believe when it wants to believe. A large part of the success of the bunco operator stems from the fact that the victim is so capable of self-delusion. He wants to believe in Santa Claus. He wants to believe he is deserving of immediate wealth, of reward, and in his mind he sees nothing illogical in his acquiring the so-called easy buck. If this yearning for quick money were not such a part of our mental makeup, there would be no bunco schemes or bunco operators.

Gold-sale bunco schemes range all the way from the crude yellow-painted brick to shares of stock in nonexistent gold mines. A common bunco scheme in the midwest of the nineteenth century was to "salt" a mine with bits of gold, have an invited visitor "find" them, and then invite him to share in the phenomenal luck of this strike by buying shares in the mine.

There have been, and there still are, and there possibly always will be, bunco gold schemes of varying degrees of sophistication. People still buy shares in mines they have never seen and have never investigated, simply on the word of a glib con man and the allure of a colorful, impressive, well-printed prospectus.

In one recent bunco, eight cans, each holding three gallons, were deposited with an armored car company. Each of the cans was supposed to contain a liquid gold concentrate having a value of about a half-million dollars. The owners of the concentrate then tried to establish a line of unlimited credit with a local bank purely on the basis of this collateral. In one instance, someone had the common sense to insist that the gold concentrate be tested. Of course, it was found to be worthless. But what is even more interesting is that this scheme was tried and did succeed. The bunco amount was close to a million dollars. This scheme, which apparently should not fool even a schoolboy, fooled some bank executives and will be tried again and will succeed again.

The basic requirement for a gold-bunco scheme is a good front man, a convincing and glib talker, one who can discuss the subject of gold with force and intelligence. The second con man involved in the swindle sometimes acts as the appraiser, and he too will talk and behave convincingly as an expert. All that is needed is a victim—and there are plenty of those—and some sort of scheme. The scheme need not be elaborate or even exceptionally well-planned.

A successful swindle requires that the victim pay his money without ever seeing what he buys. He may have been induced to do this through conversations, falsified letters of credit, falsified certificates of deposit, falsified assay reports, stupidity or greed, or a combination of any of these.

Not only is this bunco suitable for the sale of gold, but it is regularly done today with real estate. Newspapers carry advertising appealing to a basic human instinct: a home out in the woods, or by the sea, or up in the mountains, anywhere away from it all where a man can be master of everything he surveys, far from the grubbiness and pettiness of the daily struggle for existence. The sale of desert land, for example, is often done by elaborately printed color brochures showing swimming pools, roads, trees, shrubbery, mountains in the distance, plus promises of a pure environment. The victims may be invited to a free dinner where they are shown slides of what is purported to be the continued development of the property. Through flattery, pressure, and every other trick known by con men, the victim is persuaded to pay and to sign for land that may be under water, or completely inaccessible, or that can be reached only after days of travel by horseback.

COUPON BOOK SALES MO

One of the most common of human characteristics is that we all like to get something for nothing. And if that isn't possible, then we all like to get a bargain. It is this appeal to our instinct for bargains that makes the discount house, a legitimate business, so popular.

But there are discount schemes that aren't quite that legitimate. They may not be against the law, but they come very close to stepping across the line.

One type of discount-scheme bunco is the type based on coupon book sales. The venture depends on two factors: the sales ability of the promoter in selling a business on the idea, and his ability to peddle coupon books through itinerant door-to-door salespeople. The usual type of small businesses that are involved are beauty parlors, garages, and service stations.

The bunco operator sets up a small office that he uses as a base from which to call on established businesses. The normal pattern is to select a business that has recently been taken over by a new owner. The pitch is that the bunco operator represents an advertising agency that has formulated a unique plan to help the owner promote his newly acquired business. The owners of the business are reluctant until they learn that this will not require any financing or any cash outlay on their part. The bunco operator simply states that this is an effort on the part of the manufacturer (of a cosmetic, for example) to secure greater distribution and acceptance of his product. Naturally, the store owner will benefit and has been selected by the manufacturer as part of a good-will campaign.

At this time the con man brings out his sample coupon book and explains how it works. Each book has a specific number of coupons, usually ten. The book informs the purchaser of free services that come with a purchase.

Now comes the contract, which specifies the number of booklets to be sold and the number of coupons and the product or service to be offered on each. This is also specified in detail in the contract. The promoter is to print and distribute, at his own expense, the coupon booklets, usually less than five hundred, throughout the neighborhood.

All of this sounds quite legitimate and may actually be the basis of a valid advertising promotion. The business gets none of the proceeds of the sale of the coupon booklets. The contract specifies that all the business owner is to get from the promotion is the advertising and good will the coupon booklets will provide, and nothing more.

The promoters of this type of business are old hands at devising various schemes to guarantee the success of their program. Thus, in a beauty-parlor booklet, they will offer two free shampoos and sets, but the buyer of the booklet will have to buy and pay for five to get the second one free. Or, the booklet may offer a sample under-the-hood car examination with each battery or tire sale. The buyer of the booklet gets caught in several ways. The first is the payment for the booklet, usually as much as the promoter thinks he can get. The second is that the offer in the booklet is usually contingent upon buying a part or a service at full list price.

People who buy booklets of this kind are often entranced by the word FREE printed in large bold letters on each coupon, implying, for example, that the purchaser will get a free front-wheel realignment, oil and filter change, new parts, etc. Very few people bother to read the fine print, and the coupon-book promoter is skillful enough to make his sale and move on to the next victim. Quite often the promoter will have a crew working an entire neighborhood. By the time the purchasers of the booklet have learned how valueless their purchase really is, the promoters will have moved on to another area.

The buyer of the booklet loses a few dollars, but the small businessman who allows his company name to be used as part of this bunco, often creates enough ill will to drive him out of business. The victims do not blame their own desire for a bargain or their willingness to be duped. Instead, they will accuse the owner of the business of shady and shoddy practices.

This scheme is not illegal. Even if the bunco operators are caught, they can always prove they worked with the cooperation and consent of the business owner. They have a contract to prove it. There is nothing illegal in a booklet that uses the word FREE in large, bold letters, but promises nothing. Once again, the law cannot protect people against themselves.

ICE-CREAM SCAM

This method of separating a victim from his wallet is one that is usually worked in foreign countries, most often against men. It takes advantage of the fact that the victim has no knowledge of the language, or of the customs of the country, and is unable to communicate his need for help.

In this scam the victim is usually alone. While walking along a crowded street he is bumped into by two men, one of whom is carrying an ice-cream cone. The cone, apparently by accident, is smeared across the victim's jacket.

The two scam operators apologize profusely and, speaking the language of the victim, offer to take the victim into a nearby tailoring shop where the jacket will be cleaned at no cost. Since the scam operators seem so contrite and since they seem more than willing to pay for the cost of cleaning, the victim accompanies them. At the tailoring shop the victim surrenders his jacket and in the confusion forgets he has his wallet in it. The scam operator disappears into another room with the jacket. After waiting for a few minutes the second operator excuses himself on the grounds that he is going to see how the cleaning is coming along. Both men, of course, disappear and are never seen again.

There are variations of this scam and the victim is sometimes taken to an apartment with one of the scam operators offering to take the jacket to a nearby tailor.

Often, the victim is on a tour, has limited time. Even if he manages to get to the police with his story, he may get some sympathy but little else. And there is no way in which the victim can follow up his complaint. The police know this and act accordingly. Even if they manage to catch the bunco operators there will be no witnesses and no one will be available to present a complaint.

OBIT SCAM

This is a common scam and it has been in use for many years. The bunco artist may work alone but sometimes has the help of a young female, often a child. By following obituary columns in newspapers he determines the name and address of someone who has recently died. Visiting the home of the deceased he claims he has an order for some item of merchandise, usually a Bible. He also states he has received a dollar as a down payment, and as a measure of his good faith, offers to return it. The members of the family, though, out of respect for the memory of the deceased, usually agree to pay the balance and obtain the Bible. The balance is anything the scam operator decides on the moment, based on the appearance of the person at the door and any glimpse he may be able to get of the inside.

If the victim refuses to buy, the advance dollar isn't returned. Instead, the scam operator insists that this will be in the form of a check from the company he represents.

MAIL SCAMS

There are all sorts of mail scams. In one, the basic scheme is a work-at-home plan and there are a large number of variations. The appeal is that of making a good income, working in spare time at your own convenience at home. The work is easy and there is a ready market for the product, whatever that product may be. All the victim needs to do is to buy the basic supplies. Somehow, though, orders for the finished work never appear, or the work may not be satisfactory.

Some mail scams are so unbelievable that they are outrageous, and yet they are often successful in separating those who are trusting and gullible from their money. Beware of ads that make impossible offers: a "magic liquid" selling for $12.98 ($36.00 for the family size) that will dissolve away excess fat while you relax in a tub; or flowers or vegetables that grow without help and will cut your food costs by 90%; or golden opportunities to invest in real estate, or oil wells, or underseas exploration for sunken vessels loaded with gold.

The safest way is never to send money, particularly cash, through the mail. Of course, there are many legitimate mail-order companies and you can safely buy from them, but there are an equally large (or larger) number of companies that are scam operators. Check with your local Better Business Bureau (BBB). If you should weaken and get caught by a mail-order scam, complain to the same bureau. In one year, the BBB received 7,000,000 calls asking for information or help.

PYRAMID SCHEMES

Pyramid schemes are common. They flourish for a while, die out, and then are revived again. There are many variations, but one of the more common is the lucky letter. You receive a letter containing possibly a dozen names and addresses. You are asked to put your name at the bottom of the list and then mail a dollar or two to the name at the top of the list. In a variation of this scheme you are asked to become a distributor of cosmetic products. You pay for the privilege but then you have the opportunity of selling the franchise to two other potential distributors. They, in turn, may also sell the franchise to others. The problem, of course, is that the cosmetics company is selling franchises, but no products.

HOW TO PROTECT YOURSELF AGAINST BUNCO

The variety and number of bunco schemes are incredible. They are often disguised as legitimate business schemes and the victim is often dazzled by what appears to be a once-in-a-lifetime chance to become financially independent, or has a chance to make money with practically no effort.

Bunco operators can take advantage of you in many, many ways. They are shrewd judges of human nature and can spot your particular weakness in a matter of seconds. To protect yourself:

- Never sign anything.
- Always delay. Ask them to come back some other time.
- Advise the door-to-door operator that you want to check with your husband (or your wife), or to give you the phone number of a satisfied neighbor, or that you want to talk to your lawyer.
- Keep in mind that nobody in this wide, wide world has the slightest interest in making you rich, or famous, or beautiful, or strong, or well-educated, unless you supply them with a very good financial reason for doing so.
- Stay away from get-rich-quick schemes.
- Don't buy merchandise that is offered to you as "hot."
- An easy way to check on a door-to-door salesman is to learn if the company he claims to represent is in the Yellow Pages of the phone book, but remember that even some fly-by-night outfits manage to get themselves listed.
- Don't be flattered if a solicitor comes to your door and addresses you by name. He may have obtained this from the nameplate on your door or he may have asked a neighbor.
- Don't fall for schemes in which you are asked to buy some product for a very low price, just for advertising purposes.

- Don't agree to schemes in which your home will be used as a sales center to which all your neighbors will be invited, and for which you will receive some kind of product without charge.
- Don't buy raffles, so-called chances, or tickets from strangers who appear at your door. Tickets for benefits are usually for the sole benefit of the con man.
- Never invite a door-to-door salesman into your home. Don't allow him to work his way in. A common ploy is for the salesman to ask for a drink of water.
- Don't be taken in by the con man's apparel. He may appear in religious garb, may seem to be blind and guided by a seeing-eye dog, or may appear to be injured or wounded in some way. Some bunco operators have stories that would melt the heart of a stone statue.
- Don't be taken in by the youth or apparent innocence of the solicitor. Some con men use very young girls or very young boys as a front. These young people may not be aware of what they are doing, but they are being manipulated.
- At one time a man's home was his castle. Today it must be his fortress. The person out there on your steps may look helpless, frightened, weary, and as if he were just trying to make an honest dollar. Given the opportunity, he may deprive you of your money and possibly your life as well.

All of this may sound harsh and perhaps it is. It is also possible that you may turn away worthy people, people who may deserve your help and consideration. Remember, though, there are many organizations devoted to helping such people, and if you do want to help, you can do no better than to make your contribution directly. If someone does ask you for help, you can do so by referring them to an established organization.

- Don't be fooled by badges, identity cards, or letters from companies. All of these can be faked or forged, and often are.
- Many bunco door-to-door operators never sell anything. You will be asked to help someone who is working his way through college, or to contribute to some worthy cause. Quite often the opening ploy of the con man is to ask you some questions, to get you to talk to him. If you refuse to "play the game," or if you make it quite clear that you don't intend to buy or to donate, he may make an attempt to make you feel guilty, or contemptible. Don't let it bother you. It's better to feel guilty than to feel like a sucker.

Chapter 10

Miscellaneous Security Information

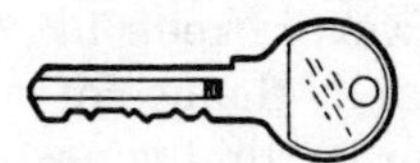

The growth of the crime rate has set the pace for a parallel growth in devices to combat crime. Combat is an unfortunate choice of word, for it implies that the law-abiding citizen and the manufacturers of security equipment aggressively seek out criminals in an effort to eliminate them. That may be true for a police department, and even then not for the entire police force, but it is not true of the citizen. It is the thief who is active and aggressive, while the average person protects himself only passively. There are some who carry guns, either licensed by the police or acting independently, who repel criminal activities with a show of force. Criminals have an option. They can use force, or not, as they wish. A citizen, though, is not in a position to counter force with force, for a number of reasons. Unlike the criminals, law-abiding citizens are not force oriented. They do not think or act in terms of force, and never have occasion to use it. You may not be permitted to carry a gun, for which a license is often required, nor are you accustomed to the use of a gun.

Finally, the criminal is protected by law in the amount of force you may use against him. You may, for example, be held up in the street by someone who looks to be about 20, but who is later shown to be only 15, with what appears to be a gun, but which is subsequently exhibited in court as a poorly made toy imitation. If your response to the attempted holdup is violence on your part that sends the young thief to the hospital, you may very well be in more legal trouble than your assailant.

There is no way in which a citizen knows what he should or should not do. In many states you are under an obligation to avoid violence if you possibly can. And you must show that you tried to retreat, that you did not seek the encounter, but that you did everything you reasonably could to avoid it. You may even be put in a position of attempting to convince a jury that the force you used in protecting yourself during a holdup was reasonable. And if, in a courtroom, the attorney for the thief can claim you overreacted, then you may find yourself in more trouble than if you had simply handed over your wallet and other possessions at once.

This doesn't mean that if someone strikes at you with a knife that all you can do is utter a verbal protest. However, keep in mind that if you succeed in overcoming the criminal by the use of force, you may be put in the position of having to prove that your defense was justifiable. Thus, if you and your wife are subjected to a string of obscenities, and you react with some sort of physical attack, you may very well find yourself the defendant in a law case.

If this sounds unfair to you, if you feel that our criminals are coddled and overprotected, the answer is not to blame the police. They simply enforce the laws enacted by legislators whom you elect. In many cases the police do excellent work in bringing criminals in to court, only to find that the criminal is given a sentence in no way commensurate with his crime, and often released so quickly that he is encouraged to continue a life of crime. What is happening is the result of a permissive society. To change it you must let your legislators know, and know emphatically, that you not only want change but that you insist on it. When legislators finally pass laws that protect the citizen rather than the criminal, it will only be because you demanded it—not as a favor, but as your right. And make no mistake about it. You have a right to be secure in the street, in a bus, train, or plane, in a railroad or airplane terminal, in a motel or hotel, in your home, or in your automobile.

The thief is not a taxpayer. You are. Everything you see around you is paid for by your work and that not only includes your home, but the streets you walk on or the highways you ride on.

THE THIEF AND TAXES

The U.S. Government can tax illegal income. This means that thieves, muggers, robbers, purse snatchers, pickpockets, or bunco artists must report their earnings and failure to do so puts them in violation of our tax laws in addition to all the other laws that govern their offenses. However, in the unlikely event that the thief does report taxable income, it is then permitted to deduct all the legitimate expenses incurred in conducting this "business."

SOCIAL SECURITY AND THE CRIMINAL

A convicted criminal cannot violate the law and then expect to get Social Security benefits on retirement. Social Security law prohibits any benefit payments to a person convicted of murder, or any felony, even though the convicted person has made payments to the Social Security fund.

INDOOR CRIME VERSUS OUTDOOR CRIME

Crimes committed indoors are often more violent than those outdoors. With an indoor crime, the criminal feels that the victim is at his mercy and may become violent, even if not provoked. Some criminals derive a sense of satisfaction in physically humiliating their victims and will do so when they have the opportunity.

SUPERMARKET CRIMES

You are at a disadvantage every time you go shopping in a supermarket. The thief knows there is money in your wallet or purse and watches you as you pay your supermarket bill at the checkout counter and determines then whether or not you are a worthwhile risk. The thief knows your arms may be full of packages and in that situation you will not be in a position to resist. He knows you will be undecided whether to protect your bags of food, your purse or wallet, and whether you should fight back. He does not have the disadvantage of the decision making process.

There are some easy rules to follow when doing supermarket shopping. These do not guarantee your protection or that you will not be a possible victim but, at least, they help improve the odds in your favor.

Don't carry your packages out to your car. Use a shopping cart even if you can easily carry the one or two packages you may have. You will still need to use both hands when pushing a shopping cart but you can take your hands away from the cart and not worry about dropping your packages. It is also unlikely that the thief is after your groceries. The preference is your wallet, purse, and/or jewelry.

Never carry more money that you need. Keep your credit card(s) and extra cash at home. Do not wear any more jewelry that you absolutely must. If at all possible, do your supermarket shopping with a friend.

Try to avoid shopping at night. Don't park your car in an isolated area. Instead, try to park your car as close to the doors of the supermarket as possible. If you must shop at night and park away from the supermarket, try to park under a light or as close to a light as you can get.

Have the car keys in your hand as you approach your car. Don't wait until you reach the car and then open your purse and fumble around in it. The car keys should be on a separate key ring and not on the same ring with your house keys. You should have no identification of any kind on your car keys, and be sure to have a duplicate set of car keys at home.

If, as you leave the supermarket, you think you are being followed to your car, turn around and walk back to the supermarket. The thief is dependent on the element of surprise so by changing your direction you are in effect notifying him of your alertness. If he follows you back into the supermarket, stay in the supermarket until you have located the manager or the security guard.

Some supermarkets have a service in which your packages are delivered to your car for you. This is an excellent security measure, but again, make sure you have your keys in hand.

THE SEX CRIMINAL

Just as there are various kinds of burglars, so too do we have various kinds of sex criminals. Some are men who simply expose themselves to women and often to female children. They usually select a time, place, and victim so as not to be readily apprehended. The usual claim made for exhibitionists is that they are sick, and should be treated as such, rather than as criminals. Granted. But they should not be out on the streets. They should not be out in society but in a hospital, preferably one associated with a prison.

In a crime of this kind—and it is a crime, no matter what other name may be given to it by sociologists and psychologists—little or no attention is paid to the victim, possibly because the victim has lost nothing of physical value, no money nor jewelry. But the experience can be traumatic, and can inflict emotional damage, possibly lasting a lifetime.

VOYEUR OR PEEPING TOM

The most passive type of sex criminal is the voyeur, or Peeping Tom. He likes to watch women undress. The cure is not to give him an opportunity. However, against this consider that, generally speaking, some women are inclined to be exhibitionists and men tend to be voyeurs. Men like to see women undress and women enjoy being watched. If a man, often accompanied by his wife, attends a night club he may get vicarious satisfaction out of what he sees on the stage. No criminal act is committed. The stage performers know what they are expected to do and often regard themselves as artists. There is no violence, no intrusion into someone else's privacy or life, no traumatic shock. The Peeping Tom, however, is intruding on someone else's privacy, and while he may do no outward physical damage, he can be responsible for emotional shock on the part of his victim if he is discovered. Like the exhibitionist, the voyeur may be in urgent need of psychological treatment.

RAPISTS

Of all sex criminals, the rapist is the most dangerous. He is the most dangerous, not only because of the rape itself, with the attendant possibilities of pregnancy and disease, but because he may very well terminate the rape with murder.

There is a common misconception that a woman who submits without fuss to a rapist is immune to violence. Violence is such a common accompaniment to rape that a victim can logically expect to be killed. It is the unharmed victim that is the exception, not the rule.

For a woman alone in a room, the best protection is a deadbolt lock and chain guard that requires a key. A telephone also provides security. And, of course, in this one respect a woman must be more security-conscious than a man, for the simple reason that she is more vulnerable.

The potential rapist is also aware that the law is on his side and not that of the victim. He can always claim that he was enticed, seduced, that the woman willingly opened her door and invited him in, or that she offered herself in exchange for money. The rapist also knows that many women are too ashamed to report a rape, and do not want the humiliation of the act known to their friends or their husbands. Women are also aware that their story of being raped may not be believed.

Since violence on the part of the rapist is almost inevitable, a woman should do everything within her power to discourage the rape. When approached she should scream, yell, kick, and create the maximum disturbance. If possible, she should pick up a heavy object and throw it, not at the rapist, but right through the window.

The rapist may be equipped with a knife or gun, and may threaten to use these weapons. True. But the odds are very likely that he will use them anyway on a submissive female. The best action in the case of potential rape is to do everything possible to forestall it.

Rape is a type of crime in which it is difficult to account for percentage changes. The number of rapes may increase or may decrease, apparently without reason. It is difficult for police to prevent rapes. One possible preventive for women is that they should avoid situations and places where rape and violent assault may take place, advice that is often difficult to follow. In an effort to decrease the incidence of rape, some police departments or other municipal departments now offer courses on the prevention of rape. Such courses may consist only of generalized advice about rape, while others supply physical education courses to enable the woman to do more than just behave as a passive victim.

BUMP AND RAPE

One of the methods used by a rapist is known as bump and rape. A man, or a group of men, will seek a road that is not well traveled, watching for a woman alone or a woman with another female passenger. They will then deliberately bump their car, often a stolen vehicle, into the woman's car. When she stops to inspect the damage and to exchange licenses she is then threatened with a knife or gun and is forced into the assailant's car. She is then driven to a remote spot, is often beaten, and is then raped and robbed.

There are a few precautions for women who drive alone. Never stop your car in a dark area. Never roll down your window to talk to the man whose car has hit yours. Never get out of your car onto the highway. Instead, drive to a police station and report the incident. If it is at all possible for you to read and remember the license number of the assailant's vehicle, do so. At least try to be able to describe the car. Give the police as much supportive information as you can.

PICKPOCKETS AND PURSE PICKS

Statistics indicate that one out of every twelve urbanites will be a pickpocket victim, and that the majority will be elderly men and women. In police language the action of a pickpocket is sometimes called grand theft person.

A pickpocket may work alone or in groups of two or three. The professional pickpocket is one of the highest types of mechanical criminals. He is an expert in human nature and accomplished in every trick of distraction and knowledge of human reflexes. By using psychology and planned physical reaction, he skillfully maneuvers a victim into a position in which his wallet can be easily removed.

Pickpockets operate in many different ways. The basic theory is: "You can't steal a man's money if he has his mind on it." The timing of the theft is planned on the principle of misdirection. Thus, when the victim directs his interests to activities at sporting events, or if he (or she) is getting on a bus or train, or buying food or clothing, his attention is concentrated on a specific activity rather than on awareness of his general environment. And because the victim is concentrating on some other activity, he becomes unaware of what the pickpocket is doing. It is literally impossible to concentrate fully on your wallet or pocketbook. Taking an upward step into a bus, for example, means you must be highly conscious of the movement of your feet, and whether you realize it or not, this simple physical act does require concentration, even though you may not be mentally aware of what you are doing. You may be daydreaming or thinking of something else while you step upward into the bus, yet part of you directs your mental activities, the other part the physical.

But even assuming you have a high level of wallet or pocketbook awareness, there are many ways in which the victim can be distracted. He may be accidentally burned by a cigarette. The pickpocket, in a restaurant, may accidentally spill a glass of water on him. There are amorous women pickpockets who put their arms around a victim, rub against him, and at the same time pick his pocket. The pickpocket may jostle or bump against the victim, and may do so strongly and so obviously deliberately that the first reaction of the victim is one of resentment. It does not matter. The victim has been distracted.

A pickpocket has a limited time in which to operate. He often cannot distract you more than once without arousing your suspicions. Keep your wallet where the pickpocket cannot reach it, and make it difficult for him to remove. In some instances pick-pockets will work in groups of three. They crowd the victim, and with two of them pushing and jostling, the third does his work. Again, if your wallet isn't readily available, as it would be in a hip pocket, they will move along to a more likely prospect.

BUMP AND RUN

In this pickpocket method, the thief deliberately bumps into the victim with enough force so the victim is distracted. Generally when someone bumps into us, and if the bump is obviously due to carelessness, our initial reaction is one of strong resentment. During the bump, the pickpocket lifts the victim's wallet. By the time the victim turns around the pickpocket manages to move away quickly and the victim is often not sure who jostled him. Because the victim may be angry due to being bumped, it may take him some little while to cool off and it is during this time that the pickpocket manages to disappear.

HOW TO PROTECT YOURSELF AGAINST PICKPOCKETS

1. Always try to be aware of your money or your purse while shopping, in crowds, in elevators, buses, and public places.

2. If you are jostled on a bus in a crowd, or someone stumbles into you, make sure immediately that your wallet and/or money are intact.

3. Don't display large sums of money in public.

4. Don't carry your wallet in your hip pocket in crowds, buses, etc. It is the easiest pocket to pick, even if you have your coat buttoned. When you buy a wallet, select the long, thin kind that

will hold dollar bills flat, without folding them. A bulging breast pocket is an out-and-out invitation to a pickpocket.

5. Don't put your pocketbook down on a counter, on the floor, or in a shopping cart. If you put your pocketbook down on a luncheon counter, for example, it can disappear in the single moment you turn your head away. A pickpocket or a pocketbook thief can be incredibly fast.

6. Remember not to carry, suspended over your arm, a clasp-type handbag that, when opened, will open away from your body.

7. When carrying open-type purses or basket types, don't have a wallet or money visible.

8. Don't carry loose bills—from dollar bills to those of higher denominations—or purses, or wallets in a coat or sweater pocket when shopping.

9. When carrying an armload of packages, keep your purse between your body and the packages.

10. When a stranger starts a conversation with you, stay aware of your wallet and your money. This is a very common diversionary gambit.

11. If you are approached by strangers who start conversations as to aches and pains, rheumatism, or other illness and proceed to touch you, or they tell you they are medical people, or are able to heal you by touch—watch out! They are touching your money as well.

12. Beware of approaches by individuals who claim to be from the general hospital or from Social Security and want to check you physically for a possible hike in pension. The only thing they want to hike is their bunco income.

13. Beware of kindly individuals who assist you in crossing the street, or who want to help you onto a bus, or want to help you adjust your packages.

14. A common ploy used by female pickpockets is to drop a compact, a book, a bag and wait for you, a gentleman at all times, to pick it up for her. She may be a woman pickpocket inviting you to bend over to give her a chance at your wallet as you do.

The average pickpocket usually uses an object, such as a newspaper or a coat over the arm, to hide his actions or to dispose of your wallet, when taken.

MUGGING

If you think you are being followed while walking, there is a quick way to verify your suspicions. Change your direction quickly and if the person or persons who are following you also do so, then you know you are being targeted as a possible victim. Start running and, if you can, yell at the same time. Of course, it is very difficult to run and yell at the same time. An alternative is to carry a whistle. One such whistle (Multiflex Industries, Inc.) consists of a combined key ring and bracelet, plus a police whistle that can emit a sharp, piercing sound. When out walking, you can always wear the bracelet, knowing the whistle is always at hand. The whistle is also quickly available when you use any key mounted on the key ring. Muggers generally look for compliant victims and want to take your money with the least fuss or noise.

If, after leaving your place of work at the end of the day, the shortest route home is through some darkened, deserted streets, change your walking arrangement to safer streets even if you must then walk a longer distance. Try not to walk alone. If you suspect you are being followed, walk into the nearest open store on the way and, if necessary, call the police.

Self-Protection

We now have an enormous self-defense industry and there are so many devices, gadgets, and gimmicks you can use that it is almost impossible to describe them except in general terms. Some of the defense measures do not fall into any particular category, and so do not logically fit into any of the preceding chapters.

A mugging is somewhat different from an ordinary holdup. A mugging involves physical contact between the thief and the victim. Instead of threatening force, the mugger uses it at once. A mugging is generally carried out by two or more individuals, although there are some muggers who are loners.

In a two-man mugging, one of the attackers comes up from behind and puts his arm around the victim's throat, exerting considerable pressure. With this kind of attack, the victim's natural response is to put both his hands against the arm that is choking him. During this time he is being frisked by the second attacker who doesn't ask where the wallet is, but goes right ahead looking for it, removing jewelry and rings at the same time.

A mugger can render his victim unconscious in a matter of seconds. Since the mugger is accustomed to violence and uses it as part of his operation, any reaction on the part of the victim other than absolute submission is countered by still more violence. Quite often the mugger will end the robbery with violence against the victim that is neither logical nor justified. There is no way to predict the behavior of a mugger, particularly if he is an addict.

Fighting Back

There are certain circumstances in which you may find it advisable to fight back. The first, of course, is that you must be in reasonably good physical condition. There is no point in repelling a thief successfully if the price you pay is a heart attack. But if you are attacked by someone not using a knife or gun or a lead pipe—that is, if the thief believes he can take you using just his fists—and if you are physically capable of countering, then the first thing you should do is to keep from being attacked from behind. If possible, move your back against a wall. And don't keep the fight a nice, quiet affair. That is just what your attacker wants. Scream, yell, make as much noise as you can. This will have two effects, both helpful. The first is that it will disconcert your attacker, and second, it will make him apprehensive that the racket you are making will bring help.

Remember also that you may be fighting for your life, so this is no time to be Mr. Nice Guy. Kick and kick violently. Aim for the shinbone. If your assailant is wearing a topcoat or overcoat, a kick in the groin may not hurt, particularly if you are inexperienced in making this kind of kick. But no matter what kind of coat the thief is wearing, it doesn't cover the region right above the ankles.

If you've had just one or two lessons in judo or karate, forget it. Just strike out with your hands and feet, trying to inflict the maximum amount of physical damage.

USING A GUN

It would seem that a gun would be a logical deterrent and should be included along with door and window locks, and alarm systems. And yet a gun has such serious limitations that you can get substantial arguments pro and con about its use. Many police departments (not all) are adamantly opposed to the in-house use of a gun, warning that the average individual is really not fully aware of the responsibility involved in owning a gun.

Today, the entire emphasis is on crime prevention through the use of deterrent devices. By following perimeter, area, and point protection you can have reasonable assurance of security. But unlike door and window locks, or an alarm system, a gun is intended for use in a confrontation. Alarms and window and door locks, though, are passive devices. They do not attack. A gun in the hands of a usually law-abiding citizen is not only a deterrent but a means of punishment. Under the stress of a robbery, it is possible to use more than reasonable force, and some courts, taking into consideration the age of the thief, might punish the gun user more than the burglar.

The problem with a gun is that many people just don't know how to handle one. According to the National Safety Council, about thirteen hundred people are accidentally killed by guns each year—not out in the street, or in a holdup, but in their own homes with their own guns. And this does not include all those who are wounded by such weapons.

Gun Permit

However, if you feel a gun is your best protection, the first thing to do is to find out if you are permitted by law in your area to have a gun. Visit your local police department and discuss the matter with them. If a permit is required, make sure you get one (assuming it will be issued to you) before buying a gun. Get the advice of the police department on what kind of gun to purchase, how often to practice, and where you may practice. Don't practice by yourself in some remote wooded area. You may hurt some other person or animal. Try pistol practice only if it is supervised.

Professional Burglar and Gun

Just one final word about guns. The burglar who invades your premises may or may not have a gun. If he is a professional, you can be sure he will have his exit planned before touching so much as a single item in your home. He is not belligerent and is seldom armed. He is aware that the penalty for burglary with a weapon, such as a gun, can add many years to his prison sentence, if caught.

Amateur Burglar and Gun

The second type of home burglar, now the most common type, is the youthful offender. He may be armed with a gun, or a knife, or both. Unlike the professional, he probably doesn't have his exit planned. The best thing, if you want to avoid violence, is to give him every opportunity to escape. If you corner him, he will fight, and then there is a question of whether you or he will survive a gun battle. He may very well win since he is probably more accustomed to handling a gun than the average person. Other than professionals, such as various members of a police department or security guards, most people do not take the time for gun practice once they have learned the basic rules for handling the weapon and have fired it a few times.

Drug Addict and Gun

The third type of home burglar is the addict, and there is no way in which you can know what he is going to do. As a general rule, the addict isn't violent (yes, there are exceptions) and is only interested in earning enough for his next fix.

You should also realize that if you have a gun you must make sure it is out of the reach of the other members of your family, particularly your children. Consider also that during a burglary you may not have time enough to get to your gun and that any attempt on your part to do so could provoke a violent reaction.

All this sounds antigun, but it isn't intended to be that way. A gun is dangerous. The whole point is whether you will be qualified to accept its responsibility.

PRIVATE SECURITY GUARDS

Private security guards have long been used in business, particularly in banks, but somewhat more recently private protective services have been developed for homes. Because the cost is rather high it is shared among a group of homeowners who have gathered for self-protection. The difficulty with private security guards for homes is in getting all the affected home owners to join. As a result some home owners may get the benefit of the added security with others carrying the financial burden.

Having private security guards is not an easy or simple process. Enough home owners in a neighborhood must be sufficiently interested. Someone must devote time to interviewing private guards, and to be responsible for ensuring and monitoring guard performance. It would seem that the ideal arrangement would be to recruit a company specializing in guard work, but not all such companies are worthwhile. Many are more interested in getting new business than in supervising the business they already have.

A job as a guard is one that involves long hours and low pay and as a result is attractive only to marginal employees. The guards often work for minimum pay, are hired without training and frequently have no prior experience doing guard work. You can get a company that has had years of experience, that does screen its applicants and that does have a training program, but all of these are reflected in their charges.

Hiring private guards, without using a security company, also raises the problems of insurance and responsibility. The group hiring the guard may be collectively responsible for the actions of that guard.

A private guard is not a policeman and does not have the authority or power that a policeman has. When security guards work for a private company they usually call in by telephone or radio to detail any problems they may have encountered and in turn their company notifies the police. Whether a guard carries a gun or not is determined by the company he works for.

CIVILIAN PATROLS

The civilian patrol is an economical alternative to the use of private guards or a security company. However, it is always advisable to get in touch with your local police department before making such a move to get the benefit of their advice and expert counseling. There is always the question of legal liability, the best method or methods of maintaining a patrol, who and where to call in the event of problems, and how to coordinate the activities of the patrol with those of the police.

TYPES OF PATROLS

There are a number of ways of going about getting up a group intended for mutual self-protection. One technique presently being used is to set up a mobile patrol using cars equipped with walkie-talkies.

One of the advantages of a private citizens' mobile patrol is that it is a less expensive route than using private security guards, plus the fact that the members of the citizens' patrol have a personal interest in mutual protection. Still, there will always be the question of funds for equipment, flashlights, gasoline and car use. If the area includes a number of merchants the owners of the stores may be interested in helping sponsor a mobile patrol. In some cases, the municipality will make some contribution, but money is generally difficult to get from such sources. Sometimes the powers-that-be in the municipal hierarchy will want to have a trial period and will also want to check with their police department to evaluate the benefits of a citizens' patrol. Some local politics may be involved and there is always the problem that no one's ego be fractured.

NEIGHBORHOOD WATCH

A neighborhood watch is a variation of the Civilian Patrol. In this setup, the members of the neighborhood watch report all suspicious activities to a volunteer dispatcher. The civilian patrol may use a telephone for this purpose or may be equipped with portable CB units. While the neighborhood watch is the least sophisticated of the various kinds of self-help groups, it also requires the least amount of money. It allows the use of the services of people who cannot or do not wish to become involved in a mobile patrol.

In larger cities, it is often difficult to get the residents of a neighborhood to become sufficiently community minded to form a neighborhood watch. Such a self-help group is easier to form in a small town where the residents know each other.

According to police, a neighborhood Crime Watch program plays an important part in reducing the incidence of burglaries. The Crime Watch program teaches people how to secure their homes, to look for suspicious people in their neighborhood, and how to inform the police of crimes. Quite often, interest in a Neighborhood Watch or Crime Watch program occurs after a burglary has been committed.

Burglary is a seasonal crime and occurs more often in the winter because of the greater number of hours of darkness. Burglaries frequently take place immediately prior to a holiday such as Christmas because the burglars know their targets are filled with merchandise. Don't expect too much in the way of a solution to a burglary. Usually less than 15% are ever solved.

Glossary

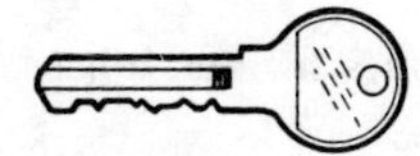

The definitions given here are necessarily brief but a more detailed explanation is given in the text. Often, the terms used in connection with security are slang expressions and may not be found in a dictionary.

A

Abortion Shake: Bunco scheme in which abortionists are threatened by phony police officers.

Acoustic Noise Generator: Device for producing broadband noise on an object.

AF: Abbreviation for audio frequency. A wave whose frequency is in the range of 20 Hz to 20 kHz (20 cycles per second to 20,000 cycles per second).

Alarm: To alert or warn. Can be used interchangeably to denote a particular warning device, such as a bell, horn, or siren, or a complete protection system.

Alarm Delay Control Switch: Switch in an alarm system that allows entry without sounding siren for specific period of time.

Amp: Abbreviation for ampere. The basic unit of electrical current.

Angle Strike: A strike in which the bolt drops vertically into slots. See also Strike.

Annunciator: Form of alarm system that sounds a buzzer or horn when a door or window is opened, but stops as soon as either is closed again.

Apartment Burglar: A burglar who specializes in apartment theft.

Area Protection: Defense of a total space. A system is said to furnish area protection when it detects and responds to an undesirable condition (such as an intruder) anywhere within the guarded space, such as an entire room or part of a room.

Armature: Moving section of a relay, corresponding to the blade of a knife switch. Mechanical portion of a relay for opening and closing electrical circuits.

Armor Collar: Metal plate placed over lock cylinder, capable of rotating to prevent removal of the cylinder by a puller.

Audio Frequency: Abbreviated as AF. A wave whose frequency is in the range of 20 Hz to 20 kHz (20 cycles per second to 20,000 cycles per second).

Automatic Light Switch: Switch that operates external floodlights when alarm system is tripped. Also used in conjunction with home alarm system to put on interior lights.

Automatic Reset: Device for automatic resetting of an alarm system after it has stopped ringing.

Automatic Telephone Dialer: Instrument connected to an alarm system for dialing a predetermined number and playing a prerecorded message in case of intrusion.

Auxiliary Lock: Door lock, not externally visible, that can be opened or closed only from inside of house or apartment.

B

Badger Game: Confidence game involving a prostitute and a male posing as her husband.

Bank Examiner Shake: Bunco involving a phony bank examiner and a depositor, often female, of the bank.

Battery Charger: Unit for recharging lead acid, alkaline or nickel-cadmium cells. When added to an alarm system that does not have an internal battery charger, will keep battery in peak operating condition.

Bookmaker Shake: Bunco scheme involving bookmakers and men posing as police officers.

Brace Lock: Steel rod that extends from the door lock with the other end fitting into a metal socket on the floor.

Breakaway Padlock: Padlock that can be opened without a key in an emergency by striking with a hammer.

Bug: Electronic device used for eavesdropping.

Bump and Run: Pickpocketing method in which the victim is bumped by the pickpocket immediately prior to lifting his wallet. The pickpocket then moves away as rapidly as possible.

Bunco: General name applied to confidence schemes.

Burglary: A crime against a place or property, not against people directly.

Buzzer: An inexpensive, not very effective type of alarm. So called because of the type of sound it produces.

C

Capacity Alarm: Alarm that is sensitive to the presence of a moving person. An alarm system which senses changes in the electrical charge on a metal plate as a person approaches.

Cat Burglar: Thief who enters occupied home or apartment at night, isn't deterred by the fact that premises are occupied. Considered most dangerous type and is frequently armed.

CB: Abbreviation for Citizens Band radio. A CB unit consists of a combined receiver/transmitter (a transceiver).

CCTV: Abbreviation for closed-circuit television.

Central Station: Specific location at which warning signals from alarm systems can be received, such as the office of a detective agency, police station, or guard office in a building complex. Alarm systems that alert guards at a central station, whether via telephone lines or radio waves, are called central station alarms. If there is no audible alarm at the protected location, such systems may be called silent alarms. At the central station the alarm signal can be in the form of a flashing light on a guard panel and/or audible alarm, such as a bell or buzzer.

Chain Guard: A door fastener making use of a small length of chain and a door fastener. Can be keyed or unkeyed. Permits limited opening of door.

Charley Bar: Rectangular length of metal, hinged at one end, used for keeping sliding patio door in its closed position.

Circuit Diagram: A drawing showing the connections of an alarm system, using symbols to represent each of the parts of that system.

Closed Circuit Television: Abbreviated as cctv. Security system consisting of video camera and monitor.

Coin Smack: A type of confidence game.

Con Man: Abbreviation for confidence man. A bunco artist.

Control Unit: Part of an alarm system that receives a signal from a sensor and turns on an alarm.

Coupon Book Sales MO: Bunco scheme involving the sale of booklets promising free merchandise or services.

CPS: Abbreviation for cycle per second. 60 cps is 60 cycles per second. This abbreviation has been replaced by the Hz (Hertz). A frequency of 60 Hz is the same as 60 cps.

Creepers: Bunco scheme operated by a pair of prostitutes or a prostitute and a helper.

Crib Job: Crime in which the criminal follows a victim to his door, pushing in as soon as the door is opened. A push-in robber.

Cylinder: The part of a lock that receives the key.

Cylinder Guard: Metal plate positioned over cylinder of lock to protect it against a burglar's cylinder puller. Sometimes called a front plate.

D

Deadbolt: Solid piece of metal, usually in rectangular shape, but can also be round or semicircular, used in a lock so arranged that it cannot be pushed back externally but only by operation of the lock itself. Generally used in quality locks for maximum security.

Deadlatch: Also known as a latch bolt. Spring latch lock that uses combination of short length of metal in various shapes as a bolt together with a spring latch.

Decoy Key: A key which will not operate the locks of any doors in your home. A false key.

Detector: A sensor. A device, usually electrical, electromechanical, or electronic that can respond to a specific physical condition or a change in conditions, and develop an electrical signal as a consequence of the condition or change. Often identified according to the type of physical condition to which it responds—i.e., sound detector, vibration detector.

Deterrent: Something that hinders, discourages, or restrains a thief. The effectiveness of a security-protection system can be expressed in terms of its deterrent value. The greater the difficulty a potential intruder may have when he tries to defeat a protective device or system, the greater its deterrent value. A deadbolt lock has greater deterrent value than a spring-latch lock.

Door Chain: Connected metal links which permit limited opening of door. Comes equipped with or without locks.

Door Cords: Fitted with special connection blocks, these cords are made with flexible two-conductor cable and are used to supply electrical connections to devices or equipment mounted on hinged access closures, such as doors or windows.

Door Switch: Key-operated electrical switch. Also known as a key or shunt switch.

Door Trip: Switch that is activated by opening a door.

Double-Cylinder Mortise Deadlock: Lock that can be keyed from inside and outside. Operates rectangular metal bolt that slides horizontally into mortised section holding strike in door jamb.

Double Lock: Two separate locks but with both using the same key.

E

Electric Eye: Photoelectric device used in alarm systems.

Electric Pencil: An electrically operated scriber used to mark metal, plastic, wood, or glass, enabling owner to put personal identification on typewriters, radios, tv sets, etc.

Electrical Switch Lock: Key-operated switch used for turning electrical circuits on and off.

Emergency Switch: Panic button used to trip an alarm system manually. Same as panic button or panic switch.

Entry-Delay Switch: Used to activate built-in electronic time-delay circuit. When in the "on" position, alarm delays for specific time. This is the normal position when occupants leave the premises. Similar time delay permits the occupants to switch the system off when they return.

F

Fail-Safe: System that alerts user in the event of equipment or circuit failure. A supervised or closed-circuit protective system is said to be fail-safe in that a cut wire, bad connection, or defective sensor will cause an alarm.

Fence: A receiver of stolen merchandise.

Field: Area of coverage of an alarm device using ultrasonic, infrared, or microwaves.

Flexible Switch: On-off switch that can be bent or twisted. Also known as a strip switch.

Forced Entry: Illegal access to a protected area by the use of mechanical force.

Fox Police Lock: Metal bar fitting a recessed holder in the floor and connected at the other end to a door lock.

Frequency: Number of hertz (cycles per second). A wave having a frequency of 5 kHz is a wave that has 5000 cycles per second.

Front Plate: Metal plate used for protecting lock cylinder against a lock puller.

Fruit: Term sometimes used to identify homosexuals.

G

Gold Sale Bunco: Confidence scheme by swindlers offering stock in nonexistent gold mine.

Grand Theft Person: Police terminology for actions of a pickpocket.

H

Hasp: Fastening device consisting of a hinged metal strap that fits over a staple. The hasp is locked with a padlock.

Hasplock: Combination hasp and lock.

Heel: Part of the shackle of a padlock.

Heist: Theft.

Hot Burglar: Burglar who prefers working in same room as victim so as to keep victim under observation.

Hotel Burglar: A burglar who specializes in theft from hotel rooms.

Hz: Abbreviation for hertz or cycles per second. A 60 Hz frequency is equivalent to 60 cycles per second.

I

IC: Abbreviation for integrated circuit. An electronic circuit etched into a small section of a semiconductor material, such as silicon.

Ice-Cream Scam: Scheme in which a victim is smeared with ice cream and then escorted to private room or tailor shop for free cleaning.

Indicator Light: Light that indicates an alarm system has been activated.

Infrared: Literally, beyond red. Invisible part of the light spectrum whose rays have longer wavelengths than visible red light. A high proportion of the energy emitted by heat lamps is in the infrared region. Infrared light, used in photoelectric alarm systems, is sometimes called black light although this expression is generally applied to ultraviolet light.

J

Jamb: The upright side of a door frame. That part of the door frame that is mortised to receive the strike of a lock.

Jamb Spreader: Any tool used to widen the distance between a door and its jamb. A jimmy is a jamb spreader.

Jiggling: Act of opening a lock with any device that will raise the tumblers in the lock.

Jimmy: Short crowbar used by burglars to force doors and windows.

K

Keeper: That part of a rim lock that receives the bolt. Comparable to the strike of a mortise type lock.

Key-In-Knob Lock: Door lock in which the key is inserted into the door knob. A combination knob and door lock.

Key-Shunt Lock: Simple key-operated switch used to temporarily shunt or bypass a protective sensor, such as a magnetic switch used on a door or similar access closure, permitting authorized entry without triggering an alarm.

Key Switch: Key-operated electrical switch. Also known as a door or shunt switch.

Keyed Travel Lock: Portable lock for securing hotel or motel doors.

kHz: Abbreviation for a thousand hertz. A frequency of 5 kHz is 5000 cycles per second.

L

Larceny: Theft. Obtaining property illegally.

Latch Bolt: Also known as a deadlatch. A spring latch lock that uses a combination of a short length of metal in various shapes as a bolt, together with a spring latch.

Lemon Game: Expert card player or poolroom shark, posing as a novice.

Lock Alarm: Lock that contains a built-in alarm.

Lock Bolt: Moving part of a lock. That part of a lock that moves into a strike or keeper.

Lock Mount: Bracket for use in autos to hinder theft of stereo or tape units.

Lock Pick: Various burglary tools for picking locks, that is, opening locks without keys.

Loiding: Technique for using a shim to open a spring latch.

M

Mace: Chemical spray in a pressurized can. The spray can disable a person temporarily.

Magnetic Switch: A switch that has two or more reedlike metallic structures that carry electrical contacts and are magnetically sensitive. Used for door and window protection.

Mail Scam: Bunco scheme in which victim is promised work at home.

Mark: Intended victim.

Marriage Bunco: Confidence scheme in which women are defrauded of money in exchange for a promise of marriage.

Mercury Tilt Switch: Switch that closes if its position is changed.

mHz: Abbreviation for a million hertz. A frequency of 10 mHz is 10,000,000 cycles per second.

Microwave: Extremely high frequency radio waves. Frequencies measured in hundreds of millions of hertz (cycles per second). At these frequencies such waves bounce back or echo from various objects. Used in some types of alarm systems.

Milliamp: Abbreviation for milliampere. Also abbreviated mA. A milliampere is a thousandth of an ampere.

Modus Operandi: Abbreviated MO. Method used by a thief.

Money Belt: Belt having a secret compartment for storing money.

Money Making Machine: A type of bunco involving a machine which appears to manufacture money.

Money Pouch: Similar to money belt but is worn beneath clothing and can be used to carry larger sums.

Monitor: To watch over. An intrusion-alarm system can be said to monitor the premises. A monitor is the viewing component of a closed-circuit television system. So, monitor means to watch but it also refers to a component.

Mortise: A hole, groove or slot cut into a door or door jamb to accommodate a lock or a strike.

Mortise Lock: A lock that is mounted in a door.

Mugger: A thief who uses personal contact on a victim.

N

N.C.: Abbreviation for normally closed. Refers to devices or circuits in which electrical contacts are kept closed under normal circumstances, opening when disturbed or activated. Relays are often described as being normally-open or normally-closed types. Magnetic switches are N.C. type sensors.

N.O.: Abbreviation for normally open. Refers to devices or circuits in which electrical contacts are kept open under normal conditions, closing when disturbed or activated. A standard electrical switch is an N.O. device.

O

Obit Scam: Confidence scheme in which bunco operators use obituary columns in newspapers to locate victims.

Omnidirectional Field: Invisible detection field that radiates from an alarm device.

On-Off Switch: Switch for turning a device to its on or active position or for deactivating it—turning it off.

P

Panic Button: Manual switch for emergency alarm. Also known as an emergency switch or panic switch.

Panic Proof: Device on a lock that permits instant opening. Also used with alarm systems for instant shutoff.

Peephole: Viewer installed in a door, generally in apartments. Allows occupants to see persons outside the door without opening it.

Peeping Tom: A voyeur. A male sex deviate who secretly watches women undress.

Pendulum Switch: A switch that works as a motion sensor. Switch operates if car or part of car is moved (i.e., trunk or door or hood opened).

Perimeter Protection: Defense of the outer edge or surface of an enclosed area. A system is said to furnish perimeter protection when it detects and responds to anyone attempting to enter the area. A fence or wall is a form of physical perimeter protection.

Personal Larceny: Theft of a purse, wallet, cash or packages from a victim. Often involves personal contact between the thief and his victim.

Photoelectric: Refers to electricity or electrical signals produced by light. Sometimes called electric eyes, photoelectric cells (or sensors) are used in some intrusion alarm systems.

Pick Men: Burglars who specialize in opening locks using various tools as picks.

Pickpocket: Thief who specializes in lifting wallets from men.

Pickpurse: Comparable to a pickpocket. Thief who specializes in opening a woman's purse.

Pictorial Diagram: A method of showing the connections of an alarm system using pictures or drawings of each of the components of that system.

Pigeon Drop: A bunco or confidence game.

Plunger Switch: Switch that has a button. Used for opening or closing electrical circuits when button is depressed or released.

Point Protection: Defense of a specific object, such as a file, cabinet, safe, or object of art. A system offers point protection when it detects and responds to anyone interfering with or attempting to touch a guarded object.

Police Whistle: A whistle intended to attract attention and capable of producing a loud, piercing tone.

Pre-Entry Alarm: Alarm system that becomes operative before break-in.

Premises: A site or location. May refer to a single room, a suite, an entire building or even a tract of real estate.

Pressure Mat: Switch that is concealed under carpets or throw rugs at entrances, near patios, or at the foot of stairs. It can also be concealed under any other object that normally sustains pressure, such as a bed.

Pry Bar: A jimmy. A short crowbar used to force entry through a door or window.

Pull Trap: An intrusion detector utilizing spring-loaded contacts plus a removable clip and actuated by a trip cord. Used for protection of garage doors, doorways, gates, driveways, and skylights.

Pushbutton Lock: A type of combination lock opened by pushing buttons in a particular sequence. Does not require a key.

Push In Robber: A thief who pushes his way into a home at the moment the victim opens the door. Same as crib job.

Push Knife: Special knife with flexible thin blade that can be inserted between a door jamb and door, or window frame and window. Used by burglars to release latches and certain types of locks.

R

Radio Frequency: Abbreviated as rf. A wave whose frequency is 100,000 Hz or higher.

Relay: Electrically operated switch. There are two main types used in security: (1) an electromagnetic relay consisting of a coil magnet and a movable spring-restrained armature, plus two or more switch contacts, and (2) a solid-state relay using semiconductor devices such as transistors and silicon-controlled rectifiers (SCRs).

Remote Alarm: Alarm device located elsewhere than on the protected premises—for example, an alarm bell that rings in the home if someone tries to enter a guest cottage elsewhere on an estate. A central station alarm is one type of remote alarm.

Resistor: Component for limiting or controlling the amount of current flowing in a circuit.

RF: Abbreviation for radio frequency. A wave whose frequency is 100,000 hertz or higher.

Rim Lock: A lock designed to be mounted on, rather than in, a door.

Ringoff: Discontinuance or cessation of alarm.

Robbery: Stealing from a person, or persons, through the use of force or intimidation.

Round-Key Lock: Lock using a round key. Helpful in security since it is difficult to duplicate the key or pick a lock that uses such a key.

Royal Shim: Playing card used as a shim.

S

Saw-Proof Bolt: Part of lock that contains a bolt with one or more hard steel rods inside it. Rods are free to rotate. If bolt is hacksawed, rotating rods prevent further progress of hacksaw blades.

Scam: General name applied to bunco or confidence schemes.

Scrambler: Device for converting telephone speech into unintelligible sounds. The speech is then unscrambled at the receiving end.

Second Story Man: Burglar who prefers working in vacant part of a house, even though other parts may be occupied.

Security: From the Latin, *securus*, freedom from care. Freedom from danger or fear. A feeling or condition of being safe.

Sensitivity Control: Adjustment of controlling range and sensitivity of an alarm system.

Sensor: A device, generally electrical, electronic, or electromechanical, that responds to specific physical conditions and initiates action or develops an electrical signal relating to those conditions.

Shackle: Looped cylindrical metal section of a padlock. The moving part of a padlock.

Shelf Life: The life of a battery when unused. The period of time during which a battery can deliver electrical power before failure occurs due to chemical deterioration.

Shim: Plastic playing card, portion of a metallic venetian blind, or any similar thin flexible material that can be inserted between a latch and door frame.

Shunt Switch: Key-operated electrical switch. Also known as a door or key switch.

Silent Alarm: System that does not produce audible alarm but sends a signal to a central station.

Single-Cylinder Mortise Deadlock: A lock that can be keyed from the outside only. Operates metal bolt that slides into strike mounted in a door jamb.

Slam Puller: Burglary tool used for opening car trunks or doors.

Slide Switch: Type of switch opened or closed by a sliding control. Used as on/off switch for alarm system or as an exit/entry alarm switch.

Snatch and Run: A technique used by thieves who specialize in snatching a woman's purse, or jewelry, and running off with it.

Solid State: Electronic circuit containing devices such as transistors, crystal diodes, and integrated circuits.

Spring Latch: Easily opened lock with bevel-edged bolt. Usually manufactured to coarse tolerances.

SPST Switch: Single-pole, single-throw switch. Simple on-off switch used in some alarm systems.

Square John: A citizen, otherwise law abiding, who buys merchandise from a stranger at a very low price. The square john may suspect the merchandise is stolen but asks no questions.

Staple: A loop of metal.

Strike: Rectangular section of metal having one or more openings to receive the bolt (or bolts) of a lock. Fits into mortised section of a door jamb.

Strip Switch: Flexible switch. A switch that can be bent or twisted.

Surveillance: To keep a close watch over a person or an object.

Switch Mat: Switch placed under a mat or rug or any other object that is subject to pressure. Used for hallway and stair protection. Also called pressure mat.

T

Tamper Switch: Electrical switch used to actuate an alarm in any attempt to manipulate or modify a piece of equipment. Often used in intrusion-detection systems to prevent anyone from gaining access to the circuitry or changing the sensitivity or operating characteristics of the equipment. Generally, a plunger type or magnetic switch mounted within the equipment's enclosure.

Test Button: Switch used for momentary testing of an alarm system.

Till Tap: One type of confidence game or bunco scheme.

Tilt Detector: Sensor that is sensitive to a change in position.

Timed Ringoff: Amount of time an alarm will stay off before being automatically reactivated.

Timer: Device for turning on lights or appliances at preselected times.

Toe: Part of the shackle of a padlock.

Transducer: Device for changing one form of energy to another. An alarm is a transducer—it changes electrical energy to sound energy. Other transducers are batteries, microphones, and speakers.

Trap Zone: Area protected by an alarm system using a wave technique, such as an ultrasonic alarm.

Trip Cord: An almost invisible or hidden string, cord or wire arranged in such a way that it will be disturbed by an intruder. Generally used with a pull trap.

U

UHF: Abbreviation for ultra high frequency. Generally a wave whose frequency is above 200 megahertz or 200,000,000 cycles per second.

Ultrasonic: Literally, "beyond that which can be heard." Physical vibrations at frequencies beyond the range of human hearing, or generally above 20,000 Hz (20 kHz). The so-called silent dog whistle that produces ultrasonic signals and ultrasonic vibrations is used in some types of intrusion-alarm systems for motion detection.

V

V: Abbreviation for volts or voltage. The volt is the basic unit of electrical pressure.

Vibration Sensor: Electromechanical device used to detect mechanical vibrations such as those caused by someone jimmying a door or window or walking across a room. Some types employ a spring pendulum.

Viewer: Peephole device mounted in door so occupant can see visitors without opening the door.

VIN: Vehicle identification number.

Voltage Surge Sensor: Sensor that will keep an alarm system from activating if the line voltage increases momentarily or surges. Eliminates false triggering of an alarm.

Voyeur: A male sex deviate who secretly watches women undress. Also known as a Peeping Tom.

W **Wall Penetration:** Ability of radiated field from alarm device to pass through a wall. Wall penetration is not achieved by ultrasonic or infrared alarm systems. Radio waves do go through walls but penetration is determined by type of radio system used and amount of metal in the wall.

Wall Sensor: Sensor attached to a wall. Touch of intruder's hand breaks a magnetic contact that sends a radio impulse to a master control unit and triggers the alarm.

Warning Device: Electrical, electromechanical, mechanical, or chemical component used to call attention to a specific condition. May be audible, as an alarm bell, buzzer, horn, or siren. It can be visual, such as flashing light. It can also be odoriferous, as in the case of special substances added to an otherwise odorless, but poisonous gas, to warn of a leak.

Wheel Lock: Used for protecting costly Mag and standard wheels against theft. Replaces one lug nut on each wheel.

Window Foil Sensor: Strip of narrow, metallic foil cemented around window and forming a continuous conductive path that is part of an electrical circuit. Detects glass breakage.

Window Jammer: Wedge-shaped device for locking windows.

Window Latch: Locking device mounted at the center of the sash of a window. Can be keyed or unkeyed. It consists of a movable portion and a fixed part called the strike.

Y **Yoke Man:** A mugger who comes up behind a victim and puts his arm around the victim's throat.

Appendix

Manufacturers

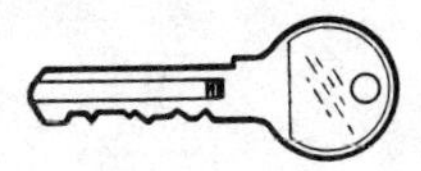

Various manufacturers are mentioned in the preceding pages, and their names, and some others, are supplied here. This is not a recommendation but this information is supplied in the event you may want to buy one or more of the products and cannot obtain them locally or if you want more specific details, including suggested retail prices and model numbers.

A. C. Custom Electronics, Inc.
686n Alpha Dr.
Highland Heights, OH 44143

All-Rite Industries, Inc.
2470 N.W. 151st St.
Opa-Locka, FL 33054

AMF/Paragon
Consumer Products Division
606 Parkway Blvd.
Two Rivers, WI 54241

Audiotronics Corp.
7428 Bellaire Ave.
North Hollywood, CA 91605

Bell System Home Consultant
200 Madison Ave.
Suite 1008
New York, NY 10016

Belwith
P.O. Box 1057
Pico Revera, CA 90660

Bolen Industries, Inc.
789 Main St.
Hackensack, NJ 07601

Burbank Enterprises, Inc.
950-D North Rengstorff Ave.
Mountain View, CA 94043

Cable Electric Products, Inc.
Box 6767
Providence, RI 02940

CCS Communication Control, Inc.
633 Third Ave.
New York, NY 10017

Dynascan Corporation
6460 Cortland St.
Chicago, IL 60635

Eastman Kodak Co.
Rochester, NY 14650

Emhart Industries, Inc.
560 Alaska Ave.
Torrance, CA 90503

General Electric Co.
1265 Boston Ave.
Bridgeport, CT 06602

Guardian Electronics, Inc.
31117 W. Via Colinas
Westlake Village, CA 91362

Hager Bolt Co.
St. Louis, MO 63104

Hydrometals, Inc.
299 Park Ave.
New York, NY 10171

Intermatic Inc.
International Plaza
Spring Grove, IL 60081

Kwikset
516-T E. Santa Ana St.
Anaheim, CA 92803

Leigh Products, Inc.
Ives Div.
New Haven, CT 06508

Lustre Line Products
Richmond & Norris Sts.
Philadelphia, PA 19125

Master Lock Co.
2600 N. 32nd St.
Milwaukee, WI 53210

Medeco Security Locks, Inc.
U.S. Route 11
W. Salem, VA 24153

Midland International Corp.
1690 North Topping
Kansas City, MO 64120

Moss Lock Mfg. Co., Inc.
7600 N.W. 69th Ave.
Miami, FL 33166

M-P Corp.
8340-T Lyndon
Detroit, MI 48238

Multiflex Industries, Inc.
9320 Augusta Rd.
W. Columbia, SC 29169

National Manufacturing Co.
Box 577
Sterling, IL 61081

Noblit Brothers & Co.
Richmond & Norris Sts.
Philadelphia, PA 19125

Nutone-Scovill
Madison & Red Bank Rds.
Cincinnati, OH 45227

Pulsafe
TMX Systems, Inc.
3152 Kashiwa St.
Torrance, CA 90505

Rebel Electronics
Box 9368
Montgomery, AL 36108

Red Thumb Home Centers
Causeway Lumber Co.
2627 S. Andrews Ave.
Ft. Lauderdale, FL 33316

Schlage Lock Co.
Rocky Mount, NC 27801

Security Research International
160 SW 12th Ave.
Dearfield Beach, FL 33441

Sentry Door Lock Guards, Inc.
114 SW 3rd Ave., Box 326
Dania, FL 33004

Sharper Image
755 Davis St.
San Francisco, CA 94111

Shelburne Co.
110 Painters Mill Rd.
Owings Mills, MD 21117

Stanley Hardware
Div. of the Stanley Works
New Britain, CT 06050

Tapeswitch Corp. of America
100-T Schmitt Blvd.
Farmingdale, NY 11735

Taylor Lock Co.
2034 W. Lippincott St.
Philadelphia, PA 19132

Throcon Controls Corp.
20 Commerce Dr.
Telford, PA 18969

Universal Security Instruments, Inc.
10324 South Dolfield Rd.
Owings Mills, MD 21117

Wright Products, Inc.
Rice Lake, WI 54868

Index

D

E

F

G

M

N

O

P